EAI/Springer Innovations in Communication and Computing

Series Editor
Imrich Chlamtac, European Alliance for Innovation
Ghent, Belgium

The impact of information technologies is creating a new world yet not fully understood. The extent and speed of economic, life style and social changes already perceived in everyday life is hard to estimate without understanding the technological driving forces behind it. This series presents contributed volumes featuring the latest research and development in the various information engineering technologies that play a key role in this process.

The range of topics, focusing primarily on communications and computing engineering include, but are not limited to, wireless networks; mobile communication; design and learning; gaming; interaction; e-health and pervasive healthcare; energy management; smart grids; internet of things; cognitive radio networks; computation; cloud computing; ubiquitous connectivity, and in mode general smart living, smart cities, Internet of Things and more. The series publishes a combination of expanded papers selected from hosted and sponsored European Alliance for Innovation (EAI) conferences that present cutting edge, global research as well as provide new perspectives on traditional related engineering fields. This content, complemented with open calls for contribution of book titles and individual chapters, together maintain Springer's and EAI's high standards of academic excellence. The audience for the books consists of researchers, industry professionals, advanced level students as well as practitioners in related fields of activity include information and communication specialists, security experts, economists, urban planners, doctors, and in general representatives in all those walks of life affected ad contributing to the information revolution.

Indexing: This series is indexed in Scopus, Ei Compendex, and zbMATH.

About EAI

EAI is a grassroots member organization initiated through cooperation between businesses, public, private and government organizations to address the global challenges of Europe's future competitiveness and link the European Research community with its counterparts around the globe. EAI reaches out to hundreds of thousands of individual subscribers on all continents and collaborates with an institutional member base including Fortune 500 companies, government organizations, and educational institutions, provide a free research and innovation platform.

Through its open free membership model EAI promotes a new research and innovation culture based on collaboration, connectivity and recognition of excellence by community.

More information about this series at http://link.springer.com/series/15427

Lucia Knapcikova • Dragan Peraković
Marko Perisa • Michal Balog
Editors

Sustainable Management of Manufacturing Systems in Industry 4.0

Springer

EAI
RESEARCH MEETS INNOVATION

Editors
Lucia Knapcikova
Dept. of Industrial Eng. & Informatics
Technical Univ Kosice, Fac of Mfg. Tech
Presov, Slovakia

Dragan Peraković
Faculty of Transport and Traffic Sciences
University of Zagreb
Zagreb, Croatia

Marko Perisa
Faculty of Transport and Traffic Sciences,
University of Zagreb
Zagreb, Croatia

Michal Balog
Faculty of Manufacturing Technologies
Technical University of Kosice
Presov, Slovakia

ISSN 2522-8595 ISSN 2522-8609 (electronic)
EAI/Springer Innovations in Communication and Computing

ISBN 978-3-030-90461-6 ISBN 978-3-030-90462-3 (eBook)
https://doi.org/10.1007/978-3-030-90462-3

This Springer imprint is published by the registered company Springer Nature Switzerland AG
The registered company address is: Gewerbestrasse 11, 6330 Cham, Switzerland

Preface

The publication *Sustainable Management of Manufacturing Systems in Industry 4.0*, provides the most comprehensive and effective exchange of information on current developments in the management of smart manufacturing systems and Industry 4.0.

The digital age has come, so computers themselves are no longer so esteemed and need to be strengthened by intelligent and autonomous systems. With digitization as the driving force of Industry 4.0, the industry focuses on machine learning, autonomous decision-making, and, in particular, the processing and evaluation of knowledge through interoperability via the Internet: IOT (Internet of Technology) and IOP – (Internet of People). The world moves further into the digital age, generating vast amounts of data and born digital content. And how will we access this content? We have to systematically develop tools to break down barriers of accessibility for future generations.

Manufacturing systems tend to be developed mostly based on lean or agile approaches to reach high level of competitiveness in the global environment.

Our priority is to create an environment that rewards excellence transparently and builds recognition objectively regardless of age, economic status or country of origin, where no membership fees or closed-door committees stand in the way of research career.

Results from different fields of science are presented in this book, but ultimately all have everything the same, they are for the people, from the people.

My thanks go to all the authors, without whom this publication would not have been written.

Presov, Slovakia — Lucia Knapčíková

Contents

Chapter 1
Trends in the Development of Transport Services Offered on Transport Exchanges During the Covid-19 Pandemic

Mateusz Kurowski and Katarzyna Huk

1 Introduction

Nowadays, transport is the area of logistics without which normal business and social activities would be impossible. It is responsible for the physical delivery of food and products for consumption. Transport is part of the distribution of goods that are essential to our life. It has a great impact on the activities of all enterprises and individual citizens. Even in the era of the pandemic that we are currently dealing with around the world, transport is the branch of the economy that should function at least to meet the basic needs of citizens of each country. The situation we have had and we are dealing with is difficult for many companies. Restrictions on the activities of economies and specific industries are one of the many factors contributing to the slowdown and decline in the number of transport services. We can consider transport activities from two perspectives: a company that has its own fleet and companies that provide transport services for a fee, and this is their basic activity. In times of economic slowdown, many factors may contribute to the increase or decrease in the number of transport transactions, as well as the total amount of transported goods.

In this chapter, the situation on the transport market will be analyzed from the point of view of companies providing transport services, as well as companies that report the need and willingness to transport their loads during the Covid-19 pandemic. The study is based on the analysis of the literature on the subject, the authors' own research carried out in 2020, the analysis of morbidity statistics, and the number of transactions and prices based on data from freight exchanges.

M. Kurowski (✉) · K. Huk
Department of Logistics and Information Systems, University of Zielona Góra, Institute of Management Science and Quality, Zielona Góra, Poland
e-mail: m.kurowski@wez.uz.zgora.pl; k.huk@wez.uz.zgora.pl

L. Knapcikova et al. (eds.), *Sustainable Management of Manufacturing Systems in Industry 4.0*, EAI/Springer Innovations in Communication and Computing,
https://doi.org/10.1007/978-3-030-90462-3_1

The aim of this chapter is to outline the trends in the development of transport services offered on transport exchanges during the Covid-19 pandemic.

2 The Role of Freight Exchanges in the Provision of Transport Services: An Overview of Opportunities and Challenges

In transport, many IT solutions are used to optimize the work of drivers, freight forwarders, as well as organization and management of the transport itself: starting from managing transportation units, working hours, through GPS and telematics, fuel control, work supervision, to systems that improve the flow of information. In this group, freight exchanges are key to the financial result and are becoming more and more popular. They are defined as "platforms for the exchange of information on offers for the transport of products, goods, materials, etc., as well as freight between market participants" [7]. Freight exchanges play an important role in the supply chains, as they determine the flow of information between entities. They are considered information brokers.

Online freight exchanges can be compared to the traditional market, but in virtual reality. It is here that contractors meet in one place and look for cooperation opportunities, and thus satisfy their needs. Transport exchanges are Internet platforms, most often in the form of a website, that act as a notice board. On the freight exchange market, one party to the contract has a load to be transported, and the other has an unoccupied transportation unit. The purpose of the exchanges is to find the best carrier and, on the other hand, the customer for transport services.

There are two types of transport exchanges in the classification of freight exchanges: open and dedicated [1]. As for the use of the first type of exchanges, there are no special restrictions. The basic requirement is to have a business activity and to purchase a subscription. Small and medium-sized transport companies and forwarders are the largest group of members of this kind of exchanges. Dedicated freight exchanges provide limited access only to designated entities and operate through the extranet system. They are only available to selected partners. This type of operation of freight exchanges ensures a lower number of offers, but potentially lower operational risk and cooperation with trusted partners.

The main feature of the freight exchange is the constant availability. Offers are available 24 hours a day, on each day of the week [7]. They use the Internet network to connect users, and it supports the wide access offers. This contributes to overcoming the geographic barrier and facilitates the gathering of a large number of users. By using freight exchanges, companies can reduce the number of empty runs, have greater access to potential customers or suppliers of transport services, and have lower transaction costs. Many manufacturing and trading companies have resigned from using the services of shipping companies, carry out these activities on their own, and have decided to use transport exchanges instead. The use of transport exchanges brings benefits such as: time savings, the possibility of negotiating with many contractors at the same time, access to thousands of offers from different countries, greater price competition [3]. These are advantages for small market players with a limited range of business contacts.

Teleroute was the first freight exchange established in France in 1985 [2] and was based on the Minitel system. Currently, there are many transport exchanges on the market, and the most popular in Europe are TimoCom, Trans.eu, Teleroute, Cargopedia, LKWonline, 123Cargo, and Wtransnet. The main features that differentiate freight exchanges among each other are the number of offers, users, and price. However, it should also be noted that these exchanges differ in specialization. In most cases, they are focused on specific countries or regions. It can be concluded that specialization and price are two of the conditions that determine the number of users and the number of offers available as a result. The network of the current market leader TimoCom includes 127,000 users which translates up to 750,000 offers per day [4].

In recent years, freight exchanges have evolved from the basic function of presenting cargo and unit offers to a highly developed tool with additional functions. Currently, they are communicators with which we can contact the contractor in real time; they are also an efficient channel of document flow. In addition, freight exchanges allow us to optimize the route and calculate costs. Through the tendering platform, companies can raise the best price for them. Moreover, in addition to traditional transport services, freight exchanges are publishing offers for renting warehouse space [8].

There is no perfect solution to verify available new customers [3]. Companies try to monitor users, but cases of abuse still happen. There are many problems, and thus challenges, occurring on the stock exchanges, which include:

- Unreliable contractors
- No possibility to collect debts
- The need for Internet access
- High level of offers' repeatability
- Subscription costs
- A large number of offers, which sometimes makes it difficult to find the most attractive ones

The biggest challenge faced by freight exchanges is the need to verify contractors and increase the security of the transaction. Many transactions are resold; that is, the company makes an offer to transport cargo for a certain price. The shipping company buys the offer and issues it again at a lower price. There can be several such intermediaries. The problem, however, is to carry out documentation between all entities of such a chain and to verify payments. In addition, it happens that the transported cargo is stolen. The management staff of transport exchanges are constantly working to increase the trust, security for their clients, and functionality of the exchanges. Representatives of TSL companies (transport-forwarding-logistics) indicate specific expectations regarding transport exchanges. They include [7]:

- Elimination of transport debtors
- Reducing the seasonality of transport services
- Drivers' recommendations
- Indicating safe parking lots
- Acquiring drivers
- Possibility to purchase cars and accessories

There are many challenges faced by freight exchanges. The most important thing now is to increase security and eliminate debtors. The boards of these organizations implement new solutions that contribute to overcoming the problems. Among them there are publishing blacklists, banning dishonest contractors, sharing statistics of trust and solvency of bidders, and supplying legal departments and debt departments within transport exchanges, which are at the full disposal of users.

3 The Economic Situation in Relation to the Covid-19 Pandemic

There are many aspects influencing the development of the economy. These can include, e.g., economic, legal, political, and social factors. As history has shown more than once, economic development is also influenced by the health of society. Currently, the year 2020 has confirmed this thesis that the epidemic may directly affect the state of the economy.

On January 7, 2020, a new virus was identified in the Chinese city of Wuhan. It has been designated as Covid-19 (or 2019-nCoV). It belongs to the same group as the MERS and SARS viruses [13]. It causes fever and cold symptoms, mainly in the respiratory tract, along with loss of taste and smell. Over time, new symptoms and forms of this virus have been discovered. However, it is dangerous for immunocompromised and elderly people, as symptoms can develop into bronchitis and/or pneumonia that cannot be treated with antibiotics. Currently, an effective drug is still being worked on. However, a vaccine against Covid-19 has already been developed by several pharmaceutical companies. From the beginning of 2021, most countries have started the vaccination process. Due to the limited capabilities, citizens are vaccinated according to the designated groups.

The World Health Organization (WHO) declared a pandemic on March 11, 2020. The WHO has stated that the best way to prevent and slow down the transmission of the virus is to be well informed about Covid-19, the disease it causes, and how it spreads. It is important to prevent infection by washing hands frequently, using an alcohol-based lotion, and not touching the face [12]. There are three main characteristics of a pandemic: the speed of its spread, the extent of its occurrence, and the duration of the phase between infection and the symptoms.

In Europe, the first cases of the disease were recorded in February 2020. The highest incidence rates in the first phase of the pandemic were observed in the period March–April 2020. In the period July–September, the number of cases decreased. Fig. 1.1 presents the daily number of new cases of Covid-19 between

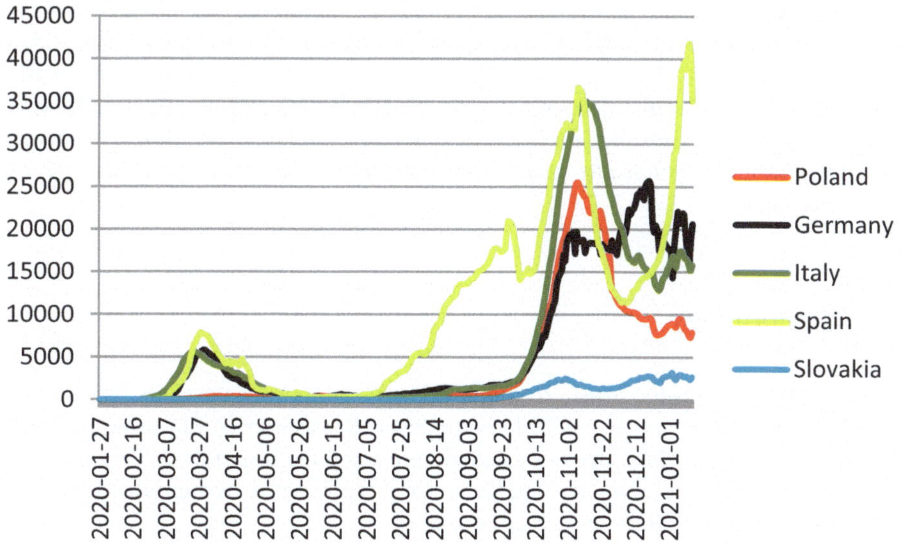

Fig. 1.1 Covid-19 daily confirmed cases in selected countries

January 27, 2020, and January 15, 2021, for Poland, Slovakia, Italy, Spain, and Germany. Counts are presented daily but they are averaged for 7-day streak.

The first cases of the disease in Europe were recorded in Italy. The number of daily new cases and its dynamics was accelerating there. Spain and Germany were the next countries with high incidence rates. In Poland and Slovakia, there is a constant tendency to increase the number of patients in the initial stage of the disease. It should be noted, however, that from October (and in the case of Spain, August) another wave of infections was recorded. There was a significant increase in infections for all of these countries, except Slovakia where the trend was relatively constant. December was another month of slowdown of the number of infections. A stable tendency at a high level has been observed in the case of Germany. Summing up, each of the countries under discussion has recorded a large increase in infections and the pandemic is still a challenge for all countries and their economies.

As the number of cases increases, the authorities of individual countries introduce restrictions and prohibitions. The key requirements for all countries are wearing protective masks, frequently disinfecting, and maintaining social distancing. Then, cultural and social facilities (cinemas, theaters, operas, museums, etc.) were closed, and the activities of restaurants, cafes and clubs, gyms, and sports clubs were limited. In addition, restrictions on movement between countries were introduced, schools and universities were closed (in time constraints), and governments recommended remote work. Limitations vary and depend on the economic, social, and cultural factors of the given community. Currently, most countries are still predicted to be in a severe crisis as a consequence of the pandemic. Just two months after the outbreak, China suffered a decrease in retail sales by 20.5%, industrial production by 13.5%, and investment by 24.5% compared to the previous year. The rate of unemployment reached 6.5% [6]. The key symptoms of the economic slowdown are: increase of inflation rate, unemployment rate, and public debt with decreasing wages at the same time. Closing selected industries and sectors of the economy also results in a deepening of the budget deficit and sovereign debt. It is also related to state subsidies for industries that have limited possibilities of activity at a given moment or have to temporarily suspend their activity by the decision of the government.

4 Methodology

This chapter presents the research conducted by the authors of the study. The survey was conducted on a sample of 200 respondents. It was carried out using an electronic questionnaire. The survey was anonymous and the respondents indicated the industry in which the company operates. They represented two groups: transport and forwarding companies and other entities.

This division was used in order to carry out a comparative analysis between transport and forwarding companies and other organizations. The survey was conducted in May 2020. The structure of the surveyed group is presented in Table 1.1.

60 transport and forwarding companies and 140 other entities took part in the study. Additionally, statistical data from the databases of transport exchanges Trans.eu and TimoCom were used. The information obtained from these databases shows:

- Demand for transport services
- Supply for transport services
- Supply/demand ratio for transport services
- Differences in freight and transport prices

The statistical data comes from the period January–May 2020 and January–December 2020, respectively. The analysis was carried out from the point of view

Table 1.1 The classification of surveyed entities

		Other companies	Transport companies	Total
Size (no. of employees)	Micro (1-9)	20	8	28
	Small (10-49)	32	24	56
	Medium (50-249)	48	16	64
	Large (>=250)	40	12	52
	Total	**140**	**60**	**200**
Routes (range)	Local	8	4	12
	Regional	8	12	20
	Country-wide	24	0	24
	International	84	36	120
	Global	16	8	24
	Total	**140**	**60**	**200**

of the situation on the transport market in Poland. Four directions have been selected to which transports from Poland are most often carried out, and where there was a large increase and dynamics of disease: Germany, Italy, Spain, and Slovakia.

The chapter presents an analysis of the demand and supply in transport and the prices of freight according to the data obtained from the two main transport exchanges operating on the European market—TimoCom [9] and Trans.eu [11]. Transport offers and freight prices for Poland were analyzed periodically. Saturdays and Sundays were not included in the study.

Based on the information on transactions carried out on freight exchanges, a summary of demand and supply for transport services was prepared. Transport has been divided into two key groups for analysis [10]:

- FTL (blue in the diagrams)—full truck loads, means transport during which the truck carries one load and is therefore fully loaded.

- LTL (red in the charts)—less than truck load, bulk transport, deadline for partial truck load. Usually, if the truck is not fully loaded, further partial loads are consolidated to make transportation profitable. Group loads are an alternative to the transport of partial loads.

In the following sections, detailed statistics on demand, supply, and charges for freight and transport services are presented.

5 The Situation on the Market of Transport Services During the Covid-19 Pandemic: Results and Discussion

The situation on the transport market changes with restrictions introduced by the government and the economic situation of individual entities. In the conducted research, as many as 88% (100% in the transport and forwarding industry) out of 200 respondents said that the pandemic had an impact on their activities. Of the respondents, 62% indicated that one of the main obstacles was organizational problems related to closing borders, 61% mentioned decrease in orders, and 48% of companies were forced to reduce jobs. In addition, the respondents indicated an increase in operating costs (30.5%) and the inability to fulfill orders for contractors (28%). However, only 1% of respondents replied that they had to terminate driver's contract. It leads to the conclusion that the pandemic crisis has affected many industries, while the TSL industry in this respect has not been too drastically affected by its effects.

One of the parameters that describe the situation in the transport market is the relation of demand to supply. Data for this part of the study was derived from TimoCom freight exchange and covers 2019 and 2020.

Higher value of the parameter indicates a greater need for transport services, which is generated by good condition of the whole economy. It means that there is a need for resources, materials, and finished goods. Companies and individuals would like to move them in order to sell or to use to add value. When the situation

Table 1.2 Freight demand/supply ratio in offers from Poland [%] [9]

From Poland to:	Year	Jan.	Feb.	Mar.	Apr.	May	Jun.	Jul.	Aug.	Sep.	Oct.	Nov.	Dec.
Germany	2019	45	50	49	44	36	48	52	40	69	69	60	52
	2020	32	41	44	18	26	51	76	74	67	75	82	66
Italy	2019	23	16	11	14	14	21	18	14	49	35	22	19
	2020	10	15	24	3	5	13	37	29	33	41	49	39
Spain	2019	9	9	7	8	10	12	39	26	23	50	25	20
	2020	5	7	7	0	3	12	25	13	45	35	16	17
Slovakia	2019	26	22	30	28	35	38	32	36	61	49	40	23
	2020	25	28	28	10	12	28	41	56	66	57	51	38

in the economy gets worse the need for physical displacement of the goods decreases too. As a result, capabilities of the supply side of the market are not used.

Table 1.2 presents demand/supply ratios calculated for offers from Poland to Germany, Italy, Spain, and Slovakia in 2019 and 2020.

The highest values of demand/supply rate are calculated for Germany [18-82%]. In all 24 analyzed months, demand for transport services in direction to Germany was the highest in the group in relation to the supply. The second highest rates were in direction to Slovakia [10-66%]. Relation of demand for transport services to supply in direction to Italy and Spain was at similar level. There was one case of no demand in April 2020 in direction from Poland to Spain. It is worth noting that in both years the highest values of the ratio were highest in the last quarter of the year. Further analysis of both years led to the conclusion that, when in 2020 there was no visible trend in demand/supply ratio, in 2019 in all four directions there was a rapid drop in April and continuous recovery after that. This observation is illustrated in Fig. 1.2.

In the beginning of 2020 demand/supply ratio for transport service in all directions was characterized by a positive trend than breakdown in the time of the first wave of the pandemic. In two months demand covered more supply than in pre-pandemic time.

When it comes to the opposite direction, from analyzed countries to Poland, the adequate data is shown in Table 1.3.

Demand/supply ratio for transport service to Poland was on average 5.33 percentage point lower than in opposite direction. In the case of freights to Poland there are no visible trends in ratio values during a year. In 2019, higher values of demand/supply ratio were calculated for Italy and Slovakia than Spain and Germany. Figure 1.3 presents the ratio values in 2020.

In contrast to operations from Poland, incoming freight demand/supply ratio values did not follow the same direction. There were no visible trends and months when ratio values representing all directions reached the bottom. The month with the lowest combined value observed was June. By the end of the year demand covered the increasing part of the supply for transport service.

By analyzing the ratio of demand to supply, these data can be refined. In the following parts of the chapter, statistical data on the demand and supply for transport

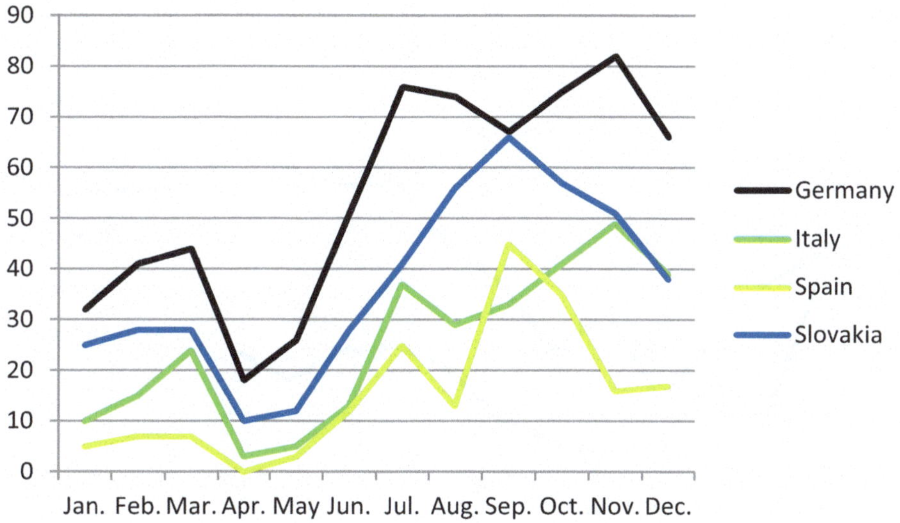

Fig. 1.2 Freight demand/supply ratio in offers from Poland [%] [9]

Table 1.3 Freight demand/supply ratio in offers to Poland in [%] [9]

To Poland from:	Year	Jan.	Feb.	Mar.	Apr.	May	Jun.	Jul.	Aug.	Sep.	Oct.	Nov.	Dec.
Germany	2019	25	6	9	18	24	19	17	22	24	17	21	20
	2020	34	9	16	5	3	5	9	18	22	21	16	21
Italy	2019	32	24	44	51	48	37	43	35	17	24	33	35
	2020	39	37	69	41	32	20	29	31	22	27	26	40
Spain	2019	41	17	24	33	36	18	8	12	3	5	26	30
	2020	35	22	36	39	37	12	5	5	0	10	31	36
Slovakia	2019	43	27	27	32	37	30	40	26	33	26	38	36
	2020	38	33	23	12	28	21	33	40	51	50	53	56

services from Poland to selected countries will be presented. The data comes from transactions carried out on the Trans.eu transport exchange, which is one of the leaders in this group of services. The data covers the period January–May 2020.

Figure 1.4 presents data on the demand for transport services from Poland to Germany. It is illustrative and has been enlarged to present the idea of the following diagrams.

As can be seen in the figure, the demand for transport was stable. The first fluctuations were recorded on March 13, 2020, and it was a drastic increase in this type of service. Then we see a decrease in transport needs, which is directly related to the restrictions introduced by the government of given countries (closing borders, limiting the activity of given industries). The situation begins to improve in mid-May, when you can see a significant increase in demand for transport services. Below, according to the same methodology, data on the demand for departures from Poland to Germany, Italy, Spain, and Slovakia are presented (Fig. 1.5).

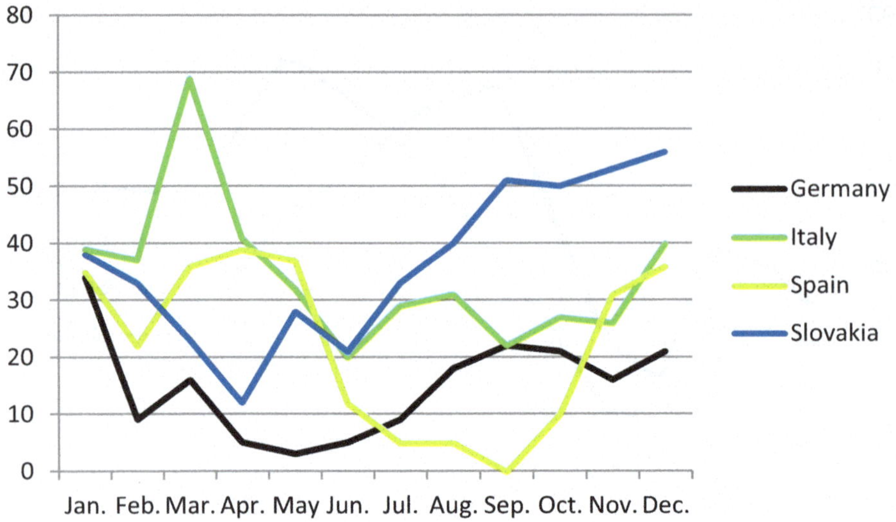

Fig. 1.3 Freight demand/supply ratio in offers to Poland in [%] [9]

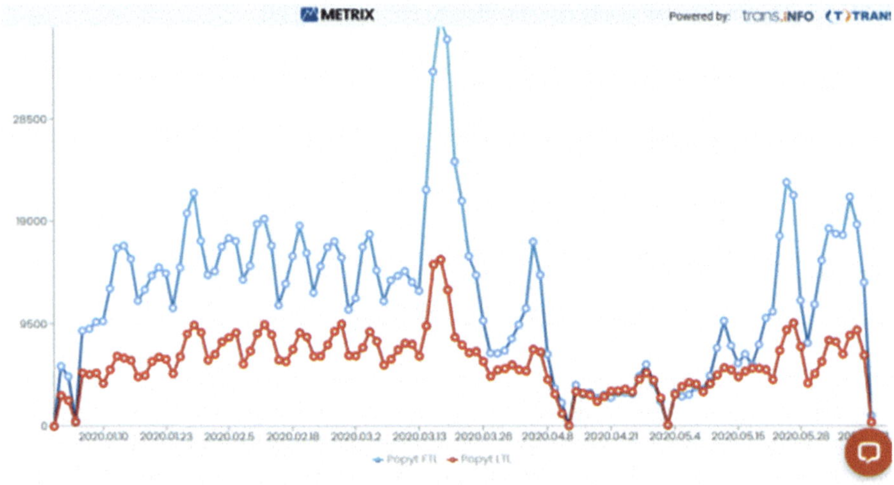

Fig. 1.4 Demand for transport services from Poland to Germany [5]

Demand for both FTL and LTL fell, but it was more pronounced in the first type of transactions. The decline began in March, the month when the pandemic began in Europe. From that point the number of new disease cases continued to grow in all described countries. In May, there was a slow increase in transport services, which is related to the growing production and exports during this period. It is a period of slow return to the pre-pandemic situation on the transport market.

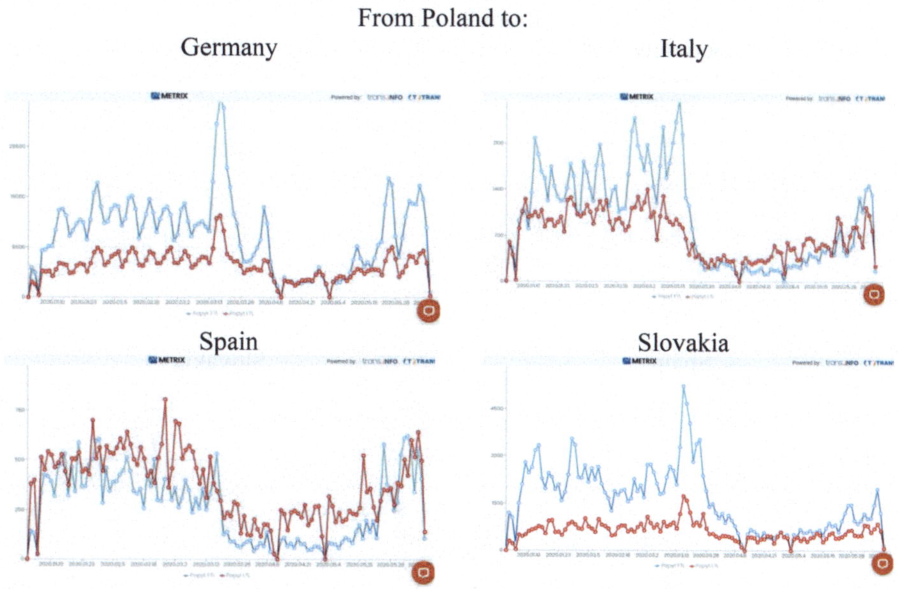

Fig. 1.5 Demand for transport services from Poland to Germany, Italy, Spain, Slovakia [5]

The next analyzed parameter was supply of transport services. Figure 1.6 shows the supply of transport from Poland to selected countries.

In mid-March 2020 there were first reported Covid-19 cases in Poland. At that time sudden drop in the supply of transport services was seen for all directions. It could have been caused by the fear of traveling and the restrictions introduced by the government. It was the first wave of an emerging pandemic when managers could not predict the upcoming changes. There was a downward trend in the supply of transport services but not as significant as in the case of demand. In some directions, the supply increased since mid-March. This was due to the search for new clients by transport and forwarding companies, which lost their regular orders at the time of the restrictions. A drop to zero means non-working days and holidays.

The analysis of transport rates was the last part of this study. These were the average prices for transport services displayed on Trans.eu transport exchange. The analysis of transport rates is the most reliable indicator that illustrates fluctuations in the transport market. It reflects the actual situation in the transport and forwarding sector and the number of transactions carried out. Supply and demand indicators on freight exchanges represent only willingness to execute a given transaction, which does not mean that the transaction was carried out. The transport rate informs about the actually performed transport service (Fig. 1.7).

The difference between the rates for FTL and LTL is evident. In addition, fluctuations in transport rates are visible at the time the pandemic started. In operations from Poland to Germany, the rates have dropped temporarily. In the case of Spain and Italy, there are large fluctuations in freight rates and at the same time the number

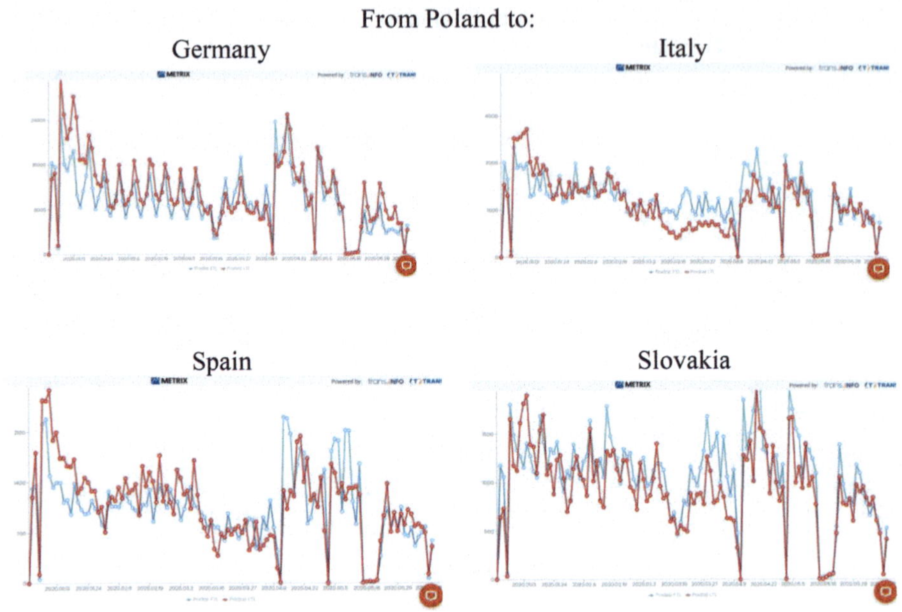

Fig. 1.6 Supply of transport services from Poland to Germany, Italy, Spain, Slovakia [5]

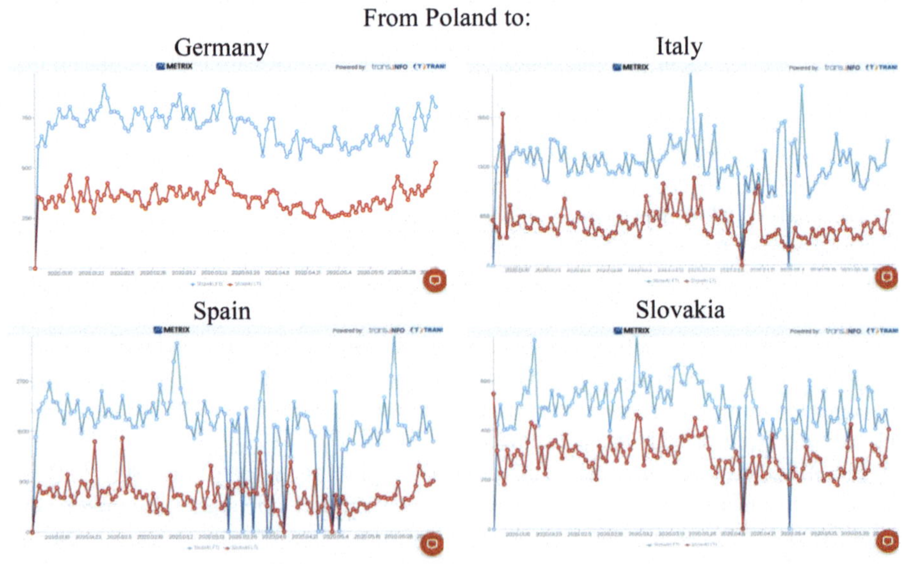

Fig. 1.7 Freight rates for transports from Poland to Germany, Italy, Spain, Slovakia [5]

of transactions. During the pandemic, there were many days when transactions to Spain were not realized at all. There is a significant difference between operations to Spain and Italy, and transport to Germany and Slovakia. In the first two countries, the number of new cases was significantly higher, and thus there were days with no operations in these directions. The above data clearly divides countries into two categories in terms of pandemic threat.

6 Summary

The Covid-19 pandemic has negatively influenced most economies around the world. It also contributes to the development or slowdown of the activity of enterprises depending on the industry or form of purchasing. In the analyzed period, for example, stationary sales in the clothing industry decreased, but e-commerce in the same industry experienced a significant increase. The functioning of transport and forwarding companies is one of the most important activities in today's economies, which is confirmed by this research. It is one of the few industries that did not experience a prolonged drop in demand and actual transactions. Summarizing research results, the following conclusions can be drawn:

- The number of freights to countries with the highest rate of new Covid-19 cases was much lower.
- There were fluctuations in the offered transport rates during the pandemic.
- The pandemic shook the transport services market, but it did not have a major impact on the reduction of drivers' workplaces or closing of the business.
- At the start of the pandemic, an increase in supply for transport was seen, which proves that most transport companies lost regular orders or contractors.
- There are no large fluctuations in the scope of transport services compared to previous periods, which proves the importance of transport in the modern economy.
- Before the outbreak of the pandemic, the managers of transport companies were struggling with the problem of finding drivers to work, and now the situation in the labor market has changed.
- Restrictions applied by the governments of all countries, especially at the border, are flexibly adjusted to improve the transport of food and the most necessary means.
- The number of transport operations has decreased due to the economic slowdown and temporary closure of many industries and facilities (catering, hotel, shopping malls, etc.).
- A greater decline was recorded for FTL than LTL.

The analyzed period was the first wave of disease and the emergence of a pandemic. At the time of publishing this study, it was known that the pandemic continued to evolve in wavelike manner. It is impossible to fully forecast and predict the situation on the market, both in transport and other services.

References

1. Kisielewski P, Leśniakiewicz M (2016) Charakterystyka i analiza porównawcza europejskich elektronicznych giełd transportowych. Logistyka 6:1368–1377
2. Lee I (2012) Electronic Commerce Management for Business Activities and Global Enterprises: Competitive Advantages. Business Science Reference, Hershey
3. Lewandowski P (2014) Freight Exchange as European Market for Transport. Logistyka 6:75–80
4. Maruszczak M (2019) Electronic freight exchanges platforms as a tool supporting the management and movement of information and loads in supply chain. In: CLC conference 2018, Tanger, Ostrava, pp 646–651
5. Metrix. https://metrix.trans.info/pages/supply. Last accessed 2020/06/13
6. Obserwator finansowy, https://www.obserwatorfinansowy.pl/tematyka/makroekonomia/trendy-gospodarcze/gdy-chiny-powracaja/. Last accessed 21 Mar 2020
7. Sosnowski J, Nowakowski Ł (2015) Elektroniczne giełdy transportowe. Difin, Warszawa
8. Starkowski D (2015) Wykorzystanie giełdy transportowej TIMOCOM Soft – und Hardware GMBH w systemie planowania operacji transportowej w przedsiębiorstwie transportowym z wykorzystaniem elektronicznego biznesu. TTS Technika Transportu Szynowego 22(12):2915–2920
9. TimoCom. https://www.timocom.pl/. Last accessed 10 May 2020
10. TimoCom lexicon. https://www.timocom.pl/lexicon/leksykon-transportowy/ltl-less-than-truck-load#lexicon. Last accessed 20 May 2020
11. Trans.eu. https://www.trans.eu/pl/. Last accessed 10 May 2020
12. WHO Coronavirus. https://www.who.int/health-topics/coronavirus#tab=tab_1. Last accessed 20 May 2020
13. WHO Homepage. https://www.who.int/csr/don/12-january-2020-novel-coronavirus-china/en/. Last accessed 28 Apr 2020

Chapter 2
Use Cases and C-ITS Communication Analysis as a Step Towards the Connected Mobility

Jiri Broz , Martin Srotyr , and Michal Jerabek

1 Introduction

Vehicle-to-vehicle communication (V2V) allows to share information about vehicle traffic (position, speed, direction of travel, braking, etc.), thanks to which potential hazards can be identified and responded to in a timely manner. V2V communication enables safe driving and, for example, avoiding traffic congestion on roads. In addition to V2V communication, vehicles are also able to communicate with transport infrastructure, such as traffic lights, via vehicle-infrastructure communication (V2I). Other functions, such as toll and parking fees, can also be communicated via V2I [1].

For V2X communication, a 5 ms transmission request is specified for a 1600 bytes message with a guaranteed delivery probability of 99.999%, with traffic controlled by events or periodically sent messages with a typical time interval of 100 ms. And in highway scenarios, communication at speeds of up to 500 km / h is envisaged [2].

In the case of autonomous vehicles, a distinction can be made between "immediate" and "remote" surroundings for communication. The immediate surroundings include vehicle-vehicle and vehicle-infrastructure communication (generally V2X); the remote surroundings include communication with information systems such as the National Traffic Information Center (in the Czech Republic NDIC), back office, or a cloud connection in general, where, e.g., parents could supervise the movement of their children in an autonomous vehicle.

While a mobile communication network is a clear choice for remote communication, two "competing" solutions dominate in near communication—a

J. Broz · M. Srotyr · M. Jerabek (✉)
Czech Technical University in Prague, Prague, Czech Republic
e-mail: brozjir9@fd.cvut.cz; srotyr@fd.cvut.cz; jerabem1@fd.cvut.cz

L. Knapcikova et al. (eds.), *Sustainable Management of Manufacturing Systems in Industry 4.0*, EAI/Springer Innovations in Communication and Computing,
https://doi.org/10.1007/978-3-030-90462-3_2

telecommunications solution based on building a network using RSUs (ITS-G5 for Europe, DSRC for America) and expanding the possibilities of base stations of a mobile communication network with direct communication between terminal devices (LTE-V = LTE-vehicle).

The choice of a suitable transmission technology is confronted with a number of requirements for system parameters, which include in particular latency, security, availability, reliability, integrity, coverage, permeability of the environment in different circumstances, etc.

Intelligent transport systems (ITS) enable traffic management. The individual service combinations must act as additional traffic management tools for the implementation of local transport policies. Effective traffic management, enabled by ITS, is a powerful tool for creating the modal shift on which most growing cities now depend [12]. Cooperative intelligent transport systems (C-ITS) enable mutual communication between ITS while V2V and V2I are important elements.

Closely related to communication, whether V2V or V2I (I2V), is the issue of security. If the vehicle is to respond to information from another vehicle or infrastructure, it is of utmost importance that the identity of the counterparty and at the same time the originality of the message are guaranteed. Testing and description of communication are given in the documents [6, 7, 9–11].

In the following part, several use cases are listed. The list contains both use cases for V2V and V2I communication. Each of the use case contains general information about the use case and some of the conditions for testing too.

The chapter with use cases is followed by the chapter of C-ITS testing in the real traffic conditions and its evaluation. This chapter describes the implemented testing of the pilot C-ITS system in the Czech Republic and the results and identified shortcomings are presented.

2 Use Cases for Testing

The following division of use cases is by criterium whether the source of information is the vehicle or the infrastructure, with a separate group of VRU—focused use cases (vulnerable road users—pedestrians, cyclists, etc.).

2.1 Use Cases with Information from Vehicle

Road Works Warning *This use case is relevant for the V2V communication as well as V2I and I2V. The use case can be split into several parts such as planned long-term static road works that can be generated in the back offices of the road operator as well as dynamic short-term road works that should be generated in the operating vehicle. The main objective is to warn drivers in advance before lane closure or other unusual situations on the road* [3].

Weather Condition Warning *This use case is related to environmental conditions on the road. The driver should get information about unexpected conditions ahead. This use case can be triggered on the infrastructure side (camera, weather stations ...) as well as by the other vehicle on the road. The source message can be generated based on the turned windshield wipers on* [4].

Stationary Vehicle *The vehicle detects that it has stopped for an undefined amount of time/has broken down and broadcasts an alert message to other vehicles. The objective of this use case is to increase the awareness of drivers about stationary vehicles to make him or her adapt his or her speed and trajectory to the situation. The other drivers around are able to avoid an accident and this use case helps to improve the safety on roads* [4].

Slow Vehicle *The vehicle detects that it goes slowly than other road users and broadcasts an alert message to other vehicles. The objective of this use case is to increase the awareness of drivers about slow-moving vehicles to make him or her adapt his or her speed and trajectory to the situation. The other drivers around are able to avoid an accident and this use case helps to improve the safety on roads* [4].

Hazardous Location Notification *A vehicle detects that it is slipping, due to a danger situation such as an animal, people, or unexpected obstacle on the road, and broadcasts an alert message to other vehicles. The objective of this use case is to increase the awareness of drivers about dangerous sections ahead to make him or her adapt his or her speed and trajectory to the situation. The alert needs to be early enough for the driver to adapt his or her speed without stress, but not too early so that the driver does not forget about the alert* [3].

Traffic Condition Warning *Sudden changes in traffic conditions downstream of the current position and in the driving direction of the vehicle may have an impact on both traffic safety and efficiency. The objective of the use case is to provide timely in-car driving assistance information on traffic conditions downstream of the current position and in the driving direction of the vehicle. The vehicle driver receives timely awareness message on the in-vehicle display* [5].

Lane Change, Merge, and Overtaking Assistance *The main objective of this use case is to provide collision risk warning for lateral maneuvers. The primary expected impact is the reduction of the risk of lateral collision and thereby improvement of traffic safety. The secondary expected impact is to smoothen lateral maneuvers and reduce any negative impact on traffic flow* [5].

Electronic Emergency Brake Lighting *A vehicle automatically detects an emergency brake and broadcasts an alert message to other vehicles. An objective of the use case is to provide relevant and fast information about the braking vehicle to other road users, and thus it leads to the safety and traffic efficiency improvements* [3].

Emergency Vehicle Approaching *The main objective of the use case is to provide information about the emergency vehicle. The emergency vehicle warns the other* *road users by the message that can be triggered by the siren or light bar. Awareness about the unclear situation can be improved. The road users can adapt their behavior and clear the road properly; this leads to the faster passing of the intersection or other road segment by the emergency vehicle* [3].

Public Transport Safety *The use case related to the public transport safety has to improve the safety in cities. The main objective of the use case is to provide information about the dangerous crossing by the public transport vehicles as well as information about the bus that is leaving the bus stop. In the future, the solution that provides information to warn before the left-turning vehicle can be applied* [4].

2.2 Use Cases with Information from Infrastructure

Wrong-Way Driving *The service is to warn the driver that he or she could stumble upon a vehicle that is driving in the wrong way. The aim is not to alert the wrong-way driver that he or she is on the wrong direction. The objective is to encourage the driver to adapt his or her speed and his or her behavior, in case of a wrong-way driving around* [3].

Public Transport Preference *Traffic lights interrupt traffic flow and therefore cause delay and emissions. For emergency, safety, environmental, traffic flow efficiency, and business reasons it may be advantageous to give priority at traffic lights to specific classes of road users* [4].

In-Vehicle Information *The road user receives (not only) speed limit notifications as he or she drives. The message subject is the dynamic speed limit given by the road manager, which is always mandatory. The road user can receive any other traffic information in the vehicle from traffic infrastructure* [3, 4].

Intersection Signal Violation *The service is to inform drivers approaching an equipped intersection that a vehicle is probably going to make a red light or stop violation. Expected benefits are to reduce the number and severity of collisions at signalized intersections* [3, 4].

Probe Vehicle Data *The service is the automatic collection of road traffic data from the vehicle to the road manager. This data can be used for real-time traffic information and management, but also to build statistical information* [4].

Railway Level Crossing *The level crossing warning message is transmitted via C-ITS messages within this service. The message describes the necessary crossing parameters and provides the approaching vehicle with information about the crossing parameters it is approaching and its current status. It is also possible to inform the driver about the status of the traffic lights at the level crossing* [4].

Green Light Optimal Speed Advisory *The service is to give drivers advices permitting to optimize their approach to a traffic light (maintain actual speed, slow down, adopt a specific speed, time to green when it is permitted by legislation)* [3].

Intersection Collision Risk Warning *ICRW application is considered as a primary road safety application. Primary road safety applications are ITS applications that target at reducing the risk of collision and thus improving the road safety. An ICRW application provides intersection collision risk warning to drivers. The warning indicates the risk of potential intersection collision risk that requires an immediate action of the driver* [7].

2.3 Use Cases of VRU

VRU-to-VRU Direct Cooperation

VRU's device embedding at least one ITS-Station (ITS-S) and potentially other types of applications [8].

Sharing Pavement Between Pedestrian and Cyclists *This is typically a VRU-to-VRU cooperation. Each VRU is equipped with an ITS-S complying with VRU standards. The VRUs are exchanging constantly standard messages enabling the detection of a risk of collision between them. When relevant, an action (in this case, an alert) can be triggered to avoid the collision* [8].

Pedestrian Crossing a Road with an E-Scooter Approaching *In this use case, one (or several) equipped VRU(s) able to receive and transmit V2X messages, i.e., in VRU-St configuration, is (are) positioned at a crossroad while an electric scooter is approaching. The e-scooter is equipped with a VRU device as well. One of the VRU ITS-S has sufficient processing capabilities to perform a risk assessment* [8].

VRU-to-Vehicle Direct Cooperation

VRU's device embedding at least one ITS-S and the vehicle is also equipped with an ITS-S [8].

Active Roadwork *By active roadwork it is meant that human workers are present and active on the roadwork zone. The VRU-to-vehicle cooperation can be achieved by VRUs using a device including an ITS-S complying with VRU standards. In such cases, the devices' ITS-S is continuously broadcasting VRU standard messages providing dynamic data elements related to their positions and movements. Vehicles are also equipped with an ITS-S complying with VRU standards and so are capable of receiving VRU standard messages and then of detecting and avoiding collision with active workers* [8].

VRU Crossing a Road *In this use case, one (or several) equipped VRU(s) able to receive and transmit V2X messages, i.e., in VRU-St configuration, are crossing a road. In the normal flow, the VRU is (are) positioned at a crossroad. The VRU standard messages are received by other vehicle ITS-S stations. In case of potential risk, the V-ITS-S broadcasts a warning message, which is received by the VRU-St* [8].

Rider Is Ejected from his or her Motorbike *A person riding a motorbike falls on a slippery road and is ejected at a certain distance from the motorbike. Approaching vehicles equipped with a V-ITS-S need to avoid running on the rider and crash the motorbike instead if the TTC is too short to brake efficiently. This use case assumes that a technical mechanism is available to pair the VRU device of the rider with the VRU device onboard of the motorbike when they are sufficiently close to each other. Such a mechanism may use existing protocols such as Bluetooth* [8].

Emergency Electronic Brake Light *The Emergency Electronic Brake Light (EEBL) application enables a vehicle to broadcast its own emergency braking situation to the surrounding vehicles, including those that have their LOS obstructed by other vehicles or bad weather such as fog or rain. In case there are multiple vehicles driving behind each other, and the first vehicle would have to perform an emergency braking, this application eliminates the delay in reaction time by subsequent vehicles: Each driver/rider is informed immediately, and collision danger could be avoided.* [8]

Motorcycle Approach Indication/Motorcycle Approach Warning *Motorcycle Approach Indication (MAI) is an application that informs a vehicle driver that an approaching motorcycle is nearby, even if the driver cannot see the motorcycle. If, based on dynamics information from both vehicles, a possible crossing with the motorcycle is detected or the relative distance between the two vehicles decreases below a given margin, an information is issued to the vehicle driver. The Motorcycle Approach Warning (MAW) application warns a vehicle driver who has a potential risk to collide with a motorcycle. This goes beyond the general notice that a motorcycle is approaching such that the MAI application provides. The MAW application is more sophisticated, because it also calculates the risk of collisions and only provides warnings to the vehicle driver if a collision is likely to occur* [8].

V2V Direct Cooperation The vehicle (with ITS-S complying with the VRU standards) detecting a hidden VRU (may be without equipping) and signaling it to other vehicles [8].

Signaling VRU Hidden by an Obstacle *Some vehicles are equipped with a front sensor (e.g., a camera) and a perception function capable of analyzing collected video and detecting VRU. When a sensor-equipped vehicle is detecting a VRU starting to cross a road, it broadcasts a standard message (e.g., DENM or CPM) signaling to other vehicles being in the C-ITS network that a hidden VRU is crossing. Receiving vehicles will act according to their relative speed and distance to the VRU* [8].

I2V Direct Cooperation A roadside equipment (RSE) (with ITS-S complying with the VRU standards) detecting a hidden VRU (may be without equipping) and signaling it to approaching vehicles (with ITS-S complying with the VRU standards) [8].

Signaled few VRUs in a Protected Area *VRUs are evolving in a protected area (e.g., pedestrian zone, roadwork, police control). The arrival of vehicles with an excessive speed is detected by means of a static or mobile roadside equipment (RSE) via its own sensors (i.e., camera). This RSE may signal the arrival of a vehicle in an excessive speed relatively to its short distance to the protected area. The RSE may also broadcast standard messages to the approaching vehicle(s) signaling the protected area. Optionally, the RSE may trigger an emergency braking at the level of a vehicle presenting a risk of collision with VRUs in the protected area* [8].

Non-equipped VRUs Crossing a Road *VRUs are non-equipped children crossing a road after leaving/boarding their scholar bus waiting for them at the bus station. Before crossing, they can be hidden by the bus itself. Vehicles intending to overtake the bus cannot perceive the hidden VRUs. A roadside equipment (RSE) senses the presence of one or several VRUs ready to cross the road and signals this risk to the vehicle (DENM or CPM) or provides them with maneuver instruction to overtake the bus when a risk of collision with a VRU does not exist anymore* [8].

VRUs Crossing at a Zebra Protected by a Traffic Light *At a crossroad, traffic lights are regulating the traffic and equipped with ITS-S. A traffic light detects the approach of a priority vehicle (e.g., a public transport) via its broadcasted CAMs. Before changing its phase from red to green, the traffic light verifies with an appropriate sensor that no VRU is engaged on the zebra crossing area. If it is the case, the traffic light phase can be changed; if not, the traffic light is waiting that all engaged VRUs have finished crossing before changing the traffic light phase* [8].

Scooter/Bicyclist Safety with Turning Vehicle *In this use case a typical critical situation is considered, where a vehicle turns right and oversees an approaching scooter or bicyclist that intended to go straight. A similar situation is considered with a vehicle approaching from the opposite direction and wants to turn left. The driver oversees the scooter or bicyclist and a collision of both road users is possible* [8].

Equipped VRU via a Third-Party Center A control center or cloud server monitoring the movement of VRUs. The VRUs may be equipped with an ITS-S detecting risks of collisions with monitored vehicles (with ITS-S complying with VRU standards) and then acting to avoid collision [8].

Network-Assisted Vulnerable Pedestrian Protection *This use case is focused on situations where a VRU is moving close to the street or crossing the street. Thanks to exchange of positioning via a global navigation satellite system (GNSS), radio-based positioning, and local sensor/camera information between users and the network via wireless communications, the network-assisted VRU protection system will determine the road user position. All this information is processed for multiple road* *users for alert generation to vehicle drivers or AD vehicles. Complementing GNSS and in-vehicle equipment with radio-based positioning is crucial in situations where GNSS reception is highly inaccurate or even impossible (tunnel, parking garage) and where in-vehicle equipment becomes unreliable because of non-line of sight (NLOS) between vehicle and VRU or bad weather conditions. Goal: To detect the presence of vulnerable road users in proximity of a vehicle with the help of the network and deliver such information to the vehicle and the VRU to avoid the potential collision with the help of accurate positioning technology* [8].

Detection of an Animal or Pedestrian on a Highway *Highways are now equipped with cameras at strategic locations, which monitor the road traffic as well as events that may happen on the highway and put vehicles at risk. Such events can be the detection of the presence of a pedestrian, or even an animal on the side of the road, likely to enter the driving lanes. These cameras are monitored in a control center, where the decision is made to trigger an alert to passing-by vehicles in a certain area, covered by a cluster of roadside units* [8].

Equipped VRU via a Third-Party RSE An RSE (with ITS-S complying with VRU standards) monitoring the movement of VRUs equipped with an ITS-S complying with VRU standards, detecting risks of collisions with monitored vehicles and then acting to avoid collision. RSE and vehicles need to be equipped [8].

Signaled Many VRUs in a Protected Area *In an urban environment, many times VRUs move in protected VRU areas and sometimes cross roads at unprotected points according to their points of interest. If many VRUs are equipped with a portable device including an ITS-S, this may create a local ad hoc network congestion problem. This can be limited if this issue is considered during the design of the overall system. Several possible approaches need to be further explored in the next parts of this standard* [8].

Intelligent Traffic Lights for All (P2I2V) *This P2I2V cooperation is achieved by each VRU using a portable device including an ITS-S complying with VRU standards. In such cases, the portable devices' ITS-S continuously broadcasts standard message providing dynamic data elements related to their positions and movements. Infrastructure (traffic lights in this case) and vehicles are also equipped with an ITS-S complying with VRU standards and so are capable of transmitting and receiving VRU standard messages and then detecting and notifying the traffic light change setting adjustment* [8].

3 C-ITS Testing in the Real Traffic Conditions and Its Evaluation

The previous section defined the individual use cases for testing. It is obvious from their description that they serve to improve transport as a complex system (increase safety, improve traffic flow, efficiency, etc.). Thanks to the participation of our university and the authors of the chapter in the pilot project C-ROADS Czech Republic, we had the opportunity to participate in the design of testing the pilot implementation of C-ITS technology in our country and then realize this testing in cooperation with other partners [13]. Testing of the C-ITS equipment took place at one of the pilot sites of the C-ROADS project and was focused specifically on the capturing data and its conformity with the following standards:

- ETSI EN 302 637-2 V1.4.1 (2019-04)
- ETSI TS 102 894-2 V1.3.1 (2018-08)

These standards specify the specific communication parameters that are contained in each C-ITS message. These standards are also known as an ITS-G5 communication that is used across the European states with dedicated frequency band 5.9GHz.

The testing of conformity with standards was realized in two main processes. The first part was focused on the communication among the cars (eventually with the equipped infrastructure) and the data capturing.

3.1 Testing and Data Capturing

These activities were performed in real traffic conditions and the following C-ITS units were used:

- Vehicle unit for logging
- Testing unit implemented in the road operator vehicle
- Roadside units along the testing road

The testing was realized on the highway D1 around the city Brno specifically between the highway exits on the 178 km and 216 km (see Fig. 2.1).

The communication data were captured all the time during the test and stored in the .PCAP format, which is commonly used in telecommunications. The data were captured by the software tool Wireshark with the plug-in[1] for the analysis of CAM, DENM, and IVIM messages that were defined by the European standards organization ETSI.

[1]The version of mentioned plug-in is available on the website http://forge.etsi.org/websvn/listing.php?repname=ITS.WIRESHARK_ITS_PLUGINS&path=%2Freleases%2Fwireshark-1.12.x%2FWindows%2F64bits%2Fdenm%2F&#a78694743a7811da4ecca34c772e9fd35

Fig. 2.1 Map with test section

For further analysis, it is necessary to note MAC addresses of the testing units and its stationIDs. These parameters of communication are required for the upcoming analysis. Without this information, it is not possible to analyze specific units in the communication log.

3.2 Log Analysis and Its Evaluation

The communication log from the testing contained high amount of packets that are irrelevant for the purposes of analysis. So, the first part of the evaluation was focused on the data processing and filtering of relevant communication packets. This part was based on the specific parameters that were noted during the test. The filtering was based on the MAC addresses of the testing equipment and the stationIDs.

The test was limited because of the lack of information about the stationID parameter from the first unit with specific MAC 00:00:000:00:00:00. This unit was generating messages that were captured on the secondary unit. It could lead to errors in the analysis. In the case of this specific testing, all of the messages (14.416 messages in the first part of testing; 14.062 messages in the second part of testing) were captured correctly in conformance with valid standards.

On the second unit, some errors were detected. This unit was also used for the capturing that means not only received packets could be analyzed but also the sent packets in the surroundings. The stationID parameter was known. So it was possible to filter data and analyze only specific packets related with the testing unit (Table 2.1).

3.3 Outputs from the Analysis

The analysis was focused on each parameter in the C-ITS message and each one was evaluated in particular. Three states could be determined:

- Error: Inconsistent with standard
- Warning: The value is not expected
- OK: In conformity with standard

In this following part, there are parameters that were detected as the error or warning. Each parameter list below also contains a description, what is wrong, and what is it stated in the standard.

semiMajorConfidence
This parameter contained in the positionConfidenceEllipse in CAM is set to the value "0," while ETSI TS 102 894-2 defines the following values:

ASN.1 representation SemiAxisLength := INTEGER{oneCentimeter(1), outOfRange (4094), unavailable(4095)} (0..4095)

Table 2.1 Number of detected errors in the C-ITS communication

Test run	Number of messages	Number of errors
1	7775	3
2	15,093	50
3	8815	5
4	12,797	44

The value "0" is admissible, but the definition states:

The value shall be set to:
1 if the accuracy is equal to or less than 1 cm,
n (n > 1 and n < 4 093) if the accuracy is equal to or less than n cm,
093 if the accuracy is equal to or less than 4 093 cm,
4 094 if the accuracy is out of range, i.e. greater than 4 093 cm,
4 095 if the accuracy information is unavailable.

The value "0" is not the statement mentioned above, so there is no exact rule, whether the value "0" is possible to use or not. For this reason, it is evaluated as warning "out of range."

semiMinorConfidence
This parameter is defined in the same way as the parameter semiMajorConfidence above.

headingConfidence
This parameter defines the quality of the related parameter headingValue. This parameter was observed with the value set to "126", which is evaluated as "out of range." This means that the quality of the parameter headingValue is irrelevant. But the headingValue is set up on "wgs84North." In this case, the value "unavailable" is expected and the message is not recommended for application.

speedConfidence
This parameter defines the quality of the related parameter speedValue. This parameter was observed with the value set to "126", which is evaluated as "Out of range." This means that the quality of the parameter speedValue is irrelevant. But the speedValue is set up on "standstill." In this case, the value "unavailable" is expected and the message is not recommended for application.

stationType
This parameter defines the type of C-ITS unit (station). Basically, it depends on its implementation and several values could be set up in the parameter. On the other hand, during the test different values were expected.

Within the analysis the value "5" (passengerCar) was expected. In some cases, the value was set up to "9" (trailer) or "11" (tram).

This abnormality could be caused by the other units in the surroundings that were detected with the same MAC address. Unfortunately, it is not possible to verify correct value. This was evaluated as the warning state.

latitude
In this parameter, the value specifying the C-ITS unit position in WGS84 coordination system is expected. Some of the packets contained value set up to "900000001," which is evaluated as unknown. This parameter is optional, so it was evaluated as the warning state.

longitude
In this parameter, the value specifying the C-ITS unit position in WGS84 coordination system is expected. Some of the packets contained value set up to "1800000001," which is evaluated as unknown. This parameter is optional, so it was evaluated as the warning state.

semiMajorOrientation
This parameter contained value set up to "3601" (unavailable). It is evaluated as warning.

altitudeValue
In this parameter, the value specifying the C-ITS unit position in WGS84 coordination system is expected. Some of the packets contained value set up to "8000001" (unavailable). This parameter is optional, so it was evaluated as the warning state.

altitudeConfidence
This parameter defines the quality of the related parameter altitudeValue. This parameter was observed with the value set to "15" (unavailable). This means that the quality of the parameter altitudeValue is "irrelevant." In this case, the value "unavailable" is expected and the message is not recommended for application.

driveDirection
This parameter specifies the driving direction. The value "0" defines the driving direction in front of the vehicle, while the value "1" means reverse direction. In some cases, the value is set up to "2" (unavailable). It is evaluated as the warning.

3.4 Resume

The real testing of C-ITS technology on the C-ROADS pilot project has shown that a specific implementation can be challenging. A number of shortcomings and non-compliance with ETSI standards were identified. The individual project partners were notified about these shortcomings. In the pilot operation these shortcomings were subsequently resolved, some even at the international level, so that they would not appear in real operation and the system could function without these problems. Testing is necessary for each phase of implementation, to verify the functionality, reliability, security, and other parameters of the entire system. It is necessary to test all implemented use cases and ideally to a very detailed level, when the use case breaks down into individual sub-processes and each of them is subjected to testing. It is also necessary to test all HW that forms the entire C-ITS system, including cross-testing HW from different manufacturers to verify mutual compatibility, which should be self-evident, but we know from testing that it is not always guaranteed.

4 Conclusion

Based on the real testing of the application of C-ITS technology in the C-ROADS project, it was found that the testing of these systems is very important and must always be implemented very consistently and carefully. Even the slightest shortcomings or differences in the technical specifications of individual providers/suppliers can lead to misunderstandings between vehicles, which can theoretically result in a dangerous event or accident. It is absolutely necessary to pay attention to the correct implementation according to the standards and subsequently verify these facts, so that the system can be relied on and can achieve the required improvement of transport—safety, traffic flow, efficiency, etc.

Currently, the driver is informed about a dangerous situation through a warning message, and its transmission to the driver (visually, acoustically, by touch, ...) must not in itself cause another dangerous situation, e.g., due to the driver's loss of attention. With the progress of the level of vehicle autonomy and communication security (concerning the guarantee of the originality of the message and the identity of the counterparty), vehicles will respond to the received messages themselves in an effort to avoid a dangerous situation or at least mitigate its consequences.

References

1. Autonomous vehicles – the legal landscape in the US and Germany (2016, July). Norton Rose Fulbright
2. Luoto P et al. (2017). Vehicle Clustering for Improving Enhanced LTE-V2X Network Performance, 17 July. http://ieeexplore.ieee.org/stamp/stamp.jsp?arnumber=7980735
3. C-ITS French Use Cases Catalog, Functional descriptions, SCOOP@F, C-ROADS FRANCE, INTERCOR (2017)
4. C-ROADS CZ Use Case catalogue 1.52
5. Dutch Profile Part A – use case catalogue (2017)

6. ETSI TS 103 191-3 V1.2.1 (2017-03) Intelligent Transport Systems (ITS); Testing; Conformance test specifications for Facilities layer protocols and communication requirements for infrastructure services; Part 3: Abstract Test Suite (ATS) and Protocol Implementation eXtra Information for Testing (PIXIT)
7. ETSI TS 101 539-2 V1.1.1 (2018-06) Intelligent Transport Systems (ITS); V2X Applications; Part 2: Intersection Collision Risk Warning (ICRW) application requirements specification
8. ETSI TR 103 300-1 V2.1.1 (2019-09) – Intelligent Transport System (ITS); Vulnerable Road Users (VRU) awareness; Part 1: Use Cases definition
9. ETSI EN 302 637-2 V1.4.1 Intelligent Transport Systems (ITS); Vehicular Communications; Basic Set of Applications; Part 2: Specification of Cooperative Awareness Basic Service
10. ETSI EN 302 637-3 V1.2.1 Intelligent Transport Systems (ITS); Vehicular Communications;Basic Set of Applications;Part 3: Specifications of Decentralized Environmental Notification Basic Service
11. ETSI TS 102 894-2 V1.3.1 Intelligent Transport Systems (ITS); Users and applications requirements; Part 2: Applications and facilities layer common data dictionary
12. European Commission (2017) C-ITS Platform, Final report Phase II
13. Lokaj Z, Srotyr M, Vanis M, Broz J (2020) Technical part of evaluation solution for cooperative vehicles within C-ROADS CZ project. Smart Cities Symposium Prague (SCSP). IEEE Press, New York. ISBN: 978-1-7281-6822-7

Chapter 3
Use of a Software Application to Calculate the Efficiency of Municipalities Using the Method of Data Envelope Analysis

Marek Jetmar, Jan Kubat, Vit Fabera, and Michal Jerabek

1 Introduction

The nature of public administration reflects the institutional aspects and principles of governing individual countries [3]. Public administration implements (fulfills) decisions of elected bodies, i.e., deals with the governance of public issues and ensures compliance of practice with law. On the European continent, its activities are carried out only on the basis and within the limits of the law; that is, its competences are strictly defined by the law. However, its constitution and organization are also subject to law. Public administration addresses the needs of society and operates on the basis of organizational structures, processes, roles, relationships, policies, and programs. It creates sustainable economic prosperity and ensures social cohesion and human well-being [1].

Public administration has very often been seen as a service to the population, a service to the public, which is reflected in the ways in which it operates [9]. Its quality affects social confidence not only in the public sector but also in the whole political system. It therefore also affects the willingness of citizens to respect the guidelines and abide by the regulation enforced by authorities, i.e., to comply voluntarily with decisions of public authorities. The quality of the institutions and the effectiveness of the services they offer play an essential role in identifying the common good, strengthening reciprocity and belonging, and enabling long-term prosperity.

M. Jetmar · J. Kubat
AMBIS, Prague, Czech Republic
e-mail: marek.jetmar@ambis.cz; jan.kubat@wo.cz

V. Fabera · M. Jerabek (✉)
Czech Technical University in Prague, Prague, Czech Republic
e-mail: fabera@fd.cvut.cz; jerabem1@fd.cvut.cz

L. Knapcikova et al. (eds.), *Sustainable Management of Manufacturing Systems in Industry 4.0*, EAI/Springer Innovations in Communication and Computing, https://doi.org/10.1007/978-3-030-90462-3_3

From the economic point of view, public administration represents a managerial structure ensuring mainly the supply of public goods. From our point of view, the allocation function of the public sector and its ability to efficiently provide public goods and public services are particularly important.

2 Materials and Methods

2.1 *Distribution of Public Services at Local Level, Czech Context*

After the territorial reform of public administration in the Czech Republic at the beginning of the millennium, the attention of the Ministry of the Interior focused on supporting processes aimed at modernizing public administration. It is about increasing the efficiency of public expenditure at all levels of public administration and effectiveness, and strengthening the transparency of decision-making and accountability to citizens. Particularly in recent years, the focus has been on implementing system approaches to quality management. Strategic Framework for the Development of Public Administration, Conceptual Document of the Ministry of the Interior for 2014–2020, adopted by the Government Resolution No. 680/2014 on the Strategic Framework for the Development of Public Administration of the Czech Republic and updated by the Government Resolution No. 1088/2016 for the period 2014-2020 [10] and the Implementation Plans including annexes, the Czech Republic, through specific objective 3.2, advocates the application of sustainable system approaches to quality management. Attention is focused not only on the institutions of state administration but also on the activities of territorial self-governing units. Services provided by the territorial public administration do not fulfill only the obligatory defined characteristics. Local and regional self-government tries to reflect within its delegated and independent competence also the requirements of stakeholders, which typically include clients of the offices, local citizens, entrepreneurs, and many others [10].

Territorial self-governing units carry out a number of development activities, including the provision of local public goods, which lead to increased satisfaction of the clients of the authorities as well as to the improvement of the quality of life for citizens in their territory. Each unit has its own procedures through which its management manages the organization—planning, organizing, communicating, or controlling the activities carried out. To ensure efficient and effective management of the organization, it is advisable to link the individual activities into a functional management system. The high number of self-governing units at the local and regional levels makes it possible to compare or develop tools to visualize the comparison.

The issue of efficiency of public services at the local level in the Czech Republic is a subject of both professional and political interest. This is due, among other things, to the high degree of atomization of local government, which is represented by 6258 independent municipalities, and the impossibility of their aggregation through administrative mergers or municipalization. This is a consequence of a high degree of autonomy, which is unusual even in the context of the European Union. Municipalities are free to exercise their autonomous powers; they are governed only by laws. The possibility of intervention by the state or state administration is limited to the supervision of legality. Municipalities defend their position very strongly and are highly sensitive to any indications of restrictions or interventions by the state.

All Czech municipalities are obliged to provide basic administrative services to a minimum extent and with regard to their population size and budget also local public services. Public libraries and elementary schools represent traditional public service provided by communities and towns. Their above-average number, compared to other countries, correlates with the high number of municipalities.

According to [7], public libraries can be defined as organizations that are established, supported, and subsidized by the society, either via local, regional, or national governments that ensure access to knowledge, information, artworks, and lifelong education via various sources and services. The main services fulfil the mission of public libraries as defined by a specific Library Act [11]; an important feature and a necessary condition of public libraries is their unlimited social-economic accessibility.

The education system provides above all education that we classify as an essential human need. It provides a wide range of knowledge, ideas, attitudes, values, and abilities for life. Education is according to paragraph 2 of the School Act [12] referred to as public service. From an economic point of view, these are net public goods, the purpose of which is to extend positive externalities. It is an important part of the public sector; schools and school facilities are nonprofit organizations of the public sector, typically primary schools [8].

In the Czech Republic, however, a consistent policy for the development of public services is not formulated by the public administration. Unlike in other countries (e.g., Germany, Austria), the availability of public services is not standardized at the territorial level. Facilities are dealt with on an ad hoc basis with a view to ensuring accessibility in the area. The competition between settlements is a significant factor.

The high number of observations makes it possible to investigate efficiency by various statistical methods. The possibility of using the DEA method for the purposes of managing the efficiency of public services is discussed by researchers.

2.2 Theoretical Background

DEA is a benchmarking tool in operations research and has a wide range of applications including but not limited to banking, business, agriculture, and transportation but also in the field of public services such as health care [8], education [5], or public libraries.

The Data Envelopment Analysis or DEA is a nonparametric method for estimating the production boundary presented in [4]. DEA measures the technical efficiency of a decision-making unit (DMU) relative to other units. Technical efficiency is therefore relative and depends on the set of all units. Units that lie at the production line are marked as effective, while units that lie below the production line are marked as ineffective. Inefficient units are also assigned an efficiency score between 0 and 1, which indicates how far the unit is from the production boundary. Units with an efficiency rate of 1 are effective and units with an efficiency rate of less than 1 are ineffective. The method was originally applied in the business sector, but later it has also been applied in evaluating the effectiveness of public services, or public administration services.

One particular issue many studies face is a heterogeneous operating environment. For DEA to make sense, however, the operating environment should be homogeneous.

3 Methodology

In the original DEA model of [4], all effective units are equal. There are many extension models of the so-called superefficiency in the literature that evaluate and compare the effective units among themselves.

We use the following assumptions when using this method. The output should be homogeneous as the quality of services is standardized by legislation. However, service providers vary widely in terms of population size, number of consumers, and amount of resources available, i.e., income of municipal budgets, etc. The consequence is that the smallest municipalities do not provide services to their citizens and use the capacities of larger municipalities. We also consider the impact of the municipality's position on development centers, or their position within the functional agglomeration, on the efficiency of the services provided. Therefore, in the first step we examine the efficiency of the whole set of municipalities providing the services. Above all, we are interested in efficiency in smaller, more homogeneous groups reflecting the size of the municipality/town and its position in the settlement structure.

The separation approach splits the heterogeneous data sample into several homogeneous subsamples according to one or more environmental variables and performs DEA separately for each subsample. The advantage of this approach is its simplicity and straightforward interpretability. However, it significantly reduces the sample size making it unusable in many studies. The all-in-one model directly includes environmental variables in DEA as inputs or outputs.

From the range of DEA models, we selected DEA model with Chebyshev distance proposed by [2] (not Euclidean, like classical DEA models) with variable returns to scale (VRS), input oriented. We choose it because of its robustness and because it is a DEA model with superefficiency so it is possible to compare also efficient units.

All the efficiency results later in the chapter are calculated by this DEA model in its linear approximation form. Anyway, according to [2], the inefficient unit order by efficiency scores of DEA model from [2] is the same as in CCR model [4] with VRS.

3.1 Variable Selection

When evaluating the efficiency of public libraries, we utilize 10 variables in total. The inputs are represented by current, operating expenditures; it means the total non-investment expenditures in CZK by the municipality on library activities (class 3314 in the sectoral classification of budget structure) in 2016 and 2017, employees, number of full-time equivalents of library employees in 2017, collection, and total number of book units owned by the library in 2016.

The outputs are represented by four variables: registrations—number of users registered in the library in 2017, circulation—number of book loans in 2017, events attendance—number of visitors of events organized by the library in 2017, and collection additions—the positive part of difference between the book collection in 2017 and 2016.

In the case of elementary schools, we considered 7 output variables and just 1 input variable for DEA and 2 segmentation variables. We were very limited by the data availability and quality.

As DEA input variables we chose, due to availability and quality of data, only:

- Current expenditures on primary schools, E(T):

The school year starts on September 1st and finishes on June 30th in the Czech Republic, but the expenditures are for the period 1.1.–31.12. So for the analyses, we transformed the expenditures as

$$E(T) = \frac{1}{3}CE(T) + \frac{2}{3}CE(T+1) \tag{3.1}$$

where E(T)	Current expenditures on primary schools for the school year T.
CE(T)	Current expenditures on primary schools for the calendar year T.

There were also separate categories for current expenditures on the 1st- and 2nd-grade primary schools, but most of the municipalities do not fill them in; although they had both grades, they just filled in the total for primary schools. So we rather do not use the split. Please note that salaries of teachers are paid from the central budget and not from municipality budgets.

As DEA output variables we considered:

- Primary schools count in a municipality, S(T)
- Total number of pupils in primary schools in a municipality, P(T)
- Total number of classes in primary schools in a municipality, C(T)
- Total number of pupils in the 1st-grade primary schools in a municipality, P1(T)
- Total number of pupils in the 2nd-grade primary schools in a municipality, P2(T)
- Total number of the 1st-grade classes in primary schools in a municipality, C1(T)
- Total number of the 2nd-grade classes in primary schools in a municipality, C2(T)

Here the 1st grade means first 5 classes (i.e., pupils from 6 to 10 years old as on September 1st, when the school year starts) and the 2nd grade comprises 4 classes (pupils aged 11–14). The output variables relate to school years.

As segmentation variables we chose

- Number of inhabitants in a municipality as of January 1, 2018, I(T). The data source is the Czech Statistical Office (CSO).
- Time to drive by car to the closest municipality with extended competence or regional development centers, centers of regional agglomerations, or metropolitan areas. The data source is the web mapping service Mapy.cz.

Public services are provided by a very heterogeneous spectrum of municipalities, represented by small villages with several dozens of citizens up to big cities. In order to make comparisons in as homogeneous groups as possible, categories of municipalities were proposed that reflect the size of the domestic population and the position of municipalities in the settlement structure. Regarding the hierarchy of settlements, it is based on the government-approved Strategy of Regional Development of the Czech Republic for 2021-2027 (Government Resolution No. 775/2019), which as higherranking settlements distinguishes regional centers and core agglomerations (essentially regional cities) and metropolitan areas (the largest cities of the Czech Republic). Most of them have the role of a municipality with extended competence (a third type of municipality, ORP). Thus, the size of the population representing potential local demand and the proximity of the municipality to the municipality with extended competence (third type) or the position of the municipality within the agglomeration or metropolitan area are monitored. The distance of the municipality to these higher centers is monitored by means of time availability when using a passenger car.

On the basis of the abovementioned data, an analysis of the effectiveness of municipalities in individual expert categories by [6] was performed. Due to the small number of observations in the cities with the largest population, these expert groups were aggregated. Prague and Brno were completely excluded as unique cases that are not comparable (Table 3.1).

The efficiency of libraries regardless of categorization by municipalities reaches an average of 0.192. Subsequently, the mean efficiency values for the given expert category were calculated. In the case of the smallest municipalities, the values range from 0.2 to 0.5, which represents a significant inefficiency. The category 101 is an outlier that showed average efficiency within a category nearly 0.8. It happened due to the fact that 25% of units were efficient within the category. This high share of

Table 3.1 Mean efficiency scores within each expert category of local public libraries in 2017

Expert category	Population (I)	Distance (D)	Units	Mean efficiency (in segments)	Mean efficiency (whole sample)
101	0–199	≤15	101	0.789	0.138
102	0–199	>15	545	0.292	0.117
201	200–499	≤15	354	0.492	0.114
202	200–499	>15	1113	0.212	0.115
301	500–999	≤15	343	0.412	0.151
302	500–999	>15	872	0.376	0.149
401	1000–1999	≤15	228	0.683	0.213
402	1000–1999	>15	484	0.583	0.208
501	2000–4999	≤15	117	0.579	0.397
502	2000–4999	>15	299	0.760	0.413
601	5000–9999	≤15	73	0.865	0.696
602	5000–9999	>15	60	0.895	0.556
701	10000–19999	–	52	0.975	0.870
801	20000–49999	–	14	1.017	0.916
901	50000–89999	–	1	2.000	2.000
1001	90000–299999	–	4	1.013	0.427

efficient units was usually caused by zero expenditures or employees and this happened more often than in other categories. It can be argued that the operation of most municipal libraries is ineffective with respect to the input and output data analyzed. As the size of the settlement increases, efficiency increases on average. The most effective are library services in towns and cities with more than 10000 inhabitants.

The influence of the agglomeration, or the position of the municipality in the hinterland of a larger city, is noticeable only from municipalities with more than 500 inhabitants. The easy accessibility of the regional centre (metropolis), where the inhabitants of the municipality spend at least a few hours during the day (e.g. employment, use of higher public services), reduces the efficiency of the operation of local libraries (Table 3.2).

Similarly, an analysis of the efficiency of primary schools, using data for 2016, shows an average increase in efficiency between categories with an increase in the population of the municipality. On average, however, schools in the highest size category of municipalities are efficient. An important finding is also the average lower efficiency of school facilities in the background of agglomerations or with good access to municipalities with extended powers or regional centers compared to municipalities with greater distance to these development centers.

A number of studies [15–18] state that as the size of the municipality or administrative unit increases, the efficiency of the provision of services increases. This

Table 3.2 Mean efficiency scores within each expert category of elementary schools in 2016

Expert category	Population (I)	Distance (D)	Units	Mean efficiency (in segments)	Mean efficiency (whole sample)
101	0–499	≤15	126	0.11	0.081
102	0–499	>15	143	0.142	0.06
301	500–999	≤15	547	0.289	0.062
302	500–999	>15	344	0.424	0.072
401	1000–1999	≤15	458	0.328	0.175
402	1000–1999	>15	240	0.431	0.199
501	2000–4999	≤15	257	0.579	0.383
502	2000–4999	>15	166	0.726	0.374
600	5000–9999	–	142	0.757	0.565
701	10000–19999	–	69	0.796	0.728
801	20000–49999	–	44	0.857	0.78
1000	50000–	–	18	1.003	0.971

was partly demonstrated in our investigation, especially in the provision of basic education services. Efficiency gains are observable when examining the efficiency of community groups vis-à-vis the whole population. At the same time, small municipalities also show an average lower efficiency within homogeneous groups.

4 Current Status of Web Application

The results of the DEA analysis can be viewed in the Web-based application in the Web browser http://ambis.fd.cvut.cz. The architecture of the application is typical for such applications: three-layer one: the database layer is composed of the Oracle DB server; application layer uses PL-SQL, R-language, and PHP; final dynamic web pages are created dynamically by PHP on the presentation layer; AJAX is used on the client side. The architecture of the application and the relational model of the database are presented in [13, 14] in more details.

The configuration on the server and the installed software are concluded in Table 3.3.

The database contains information about 6251 towns/villages in the Czech Republic (the state from 2017). The total count of towns/villages is currently 6258; that is, we have included 99.88% of them. Financial and statistical data containing budget and number of populations related to each town/village are filed in the range of 2012–2020.

After logging into the application, a table with the list of all towns/villages is displayed. It is the longest list of the application and viewing the table is the longest time operation. The load of 99% of records takes about 10 s, and the rest of the list and formatting take the next 20 s. The response of all other operations is less than

Table 3.3 Hardware and software solution for WebDEAr application

Hardware	
Processor	4 x Intel(R) Xeon(R) CPU E5-2630L v2 @ 2.40GHz
Memory	8 GB
Disk	161 + 97 GB
Software	
OS	Oracle Linux 64-bit, kernel 4.14.35
DBMS	Oracle DB Server ver. 18
Oracle Instant Client	Oracle Instant Client 19.3
R-library	R 3.3.0
WebServer	HTTPD 2.6.4
PHP	PHP 7.3

2 s. The measure was performed on the local network in one building (two segments) using Google Chrome Browser. When it was accessed using 4G mobile network, the load of 99% of records was stretched to approx. 17 s; the rest of the list and formatting take the next 20 s.

During 2020, major development work on the application was completed. After logging into the application, the signpost offers functionalities from two main modules:

(a) Information about municipalities
(b) Information about indicators

4.1 *Information About Municipalities*

This module displays basic information such as name, identification number, municipality with an authorized municipal office, municipality with extended powers, district, region, and more. At the initiative of the application guarantor, information on the type (level) of the municipality was added in 2020, which is also displayed in the indicators. The list of municipalities can be sorted according to various criteria and also filtered. For each municipality, it is possible to view more detailed information, where information on municipalities with a registry office and a building office has been added. Financial details are also available in the detail of the municipality, which can be updated and supplemented with data from the state application monitor. Financial data from 2012 are available in the application and relate to both the financial section and the financial item. In Table 3.4, the number of financial records per time period is shown.

The financial paragraphs are divided into three levels:

- Groups

Table 3.4 Number or financial records

Year/month	Number of records
2012/12	821,017
2013/12	875,486
2014/12	889,328
2015/12	848,082
2016/12	901,262
2017/12	928,670
2018/12	979,906
2019/12	966,793
2020/08	895,463

- Sections
- Subsections

At the highest level, the paragraphs are divided into seven groups:

- State security and legal protection
- Industrial and other sectors of the economy
- Revenues
- Services for the population
- Social affairs and employment policy
- General public administration and services
- Agriculture, forestry, and fisheries

4.2 Information About Indicators

There are three types of indicators in the application:

- Saved sets of efficiency calculations
- The values of selected financial paragraphs
- Comparison of efficiency of selected municipalities—own calculation

There were no changes in the application part of the stored efficiency calculation sets in 2020, but other calculated data sets were completed.

Listing the values of selected paragraph items is a newly created functionality of the application. The user must select the municipalities for which he or she wants to display data and select a specific group, section, and subsection in the tree structure or select individual paragraphs and the required time period in the drop-down list (see Fig. 3.1).

Subsequently, the data for the given selection are displayed. The data comes from the state application monitor (see above), but is transformed to work quickly in the required structure. At the database level, materialized view was used for this, which is updated every hour.

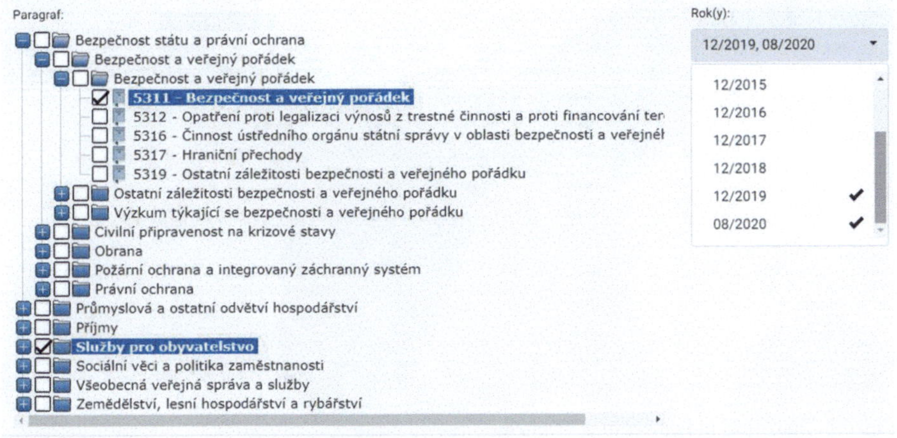

Fig. 3.1 Selection of financial paragraphs

At present, it is possible to calculate the efficiency for five basic criteria, which cover important areas of functioning of municipalities:

- Libraries
- The municipal police
- Roads
- Government
- Elementary school

For each criterion, it is possible to select a specific time period for which the calculation will be performed.

When comparing the effectiveness of its own calculation, there are three options. Either the efficiency is calculated within the categories to which the municipalities belong or the efficiency is calculated regardless of the categories, and the third option allows fictitious merging of selected municipalities and comparison of the efficiency of the resulting fictitious municipality with comparable real municipalities.

In the case of comparing efficiency within categories, the result shows information in the first column about the category of the municipality within which it is compared with other municipalities.

In addition to the result and the order of the calculated efficiency, the values used for input and output in the DEA calculation are also displayed. And in addition to these data, other statistical indicators for the given categories of municipalities are displayed—median, average, upper and lower decile of calculated efficiencies, and ranking within the region (see Fig. 3.2).

In addition to the listed functionalities of the application, it allows the management of persons/users of the application. For users with the ADMIN system role, interfaces are also available for editing data in the database, either for expanding code lists or for uploading data. User with the ANALYTIC role can see all

Kopírovat Excel Tisk Zobrazuji 10 záznamů ▾ pro řazení dle více sloupců držte při výběru klávesu SHIFT Exportovat vše do EXCELu Hledat:

Kategorie	Obec	Stupeň	Výsledek	Kategorie pořadí	Kategorie median	Kategorie průměr	Kategorie horní decil	Kategorie dolní decil	ORP	ORP pořadí	Kraj	Kraj pořadí
	Filtr:											
Si5 (500, 1.tis) obyvatel, do 15 min. dojezd	Bdeněves	I.	2,0000	1	1,5009	1,2358	2,0000	0,2935	Nýřany	1	Plzeňský kraj	1
Si5 (500, 1.tis) obyvatel, do 15 min. dojezd	Besednice	I.	2,0000	1	1,5009	1,2358	2,0000	0,2935	Kaplice	1	Jihočeský kraj	1
Si5 (500, 1.tis) obyvatel, do 15 min. dojezd	Bernartice	I.	1,5009	3	1,5009	1,2358	2,0000	0,2935	Trutnov	1	Královéhradecký kraj	1
Si5 (500, 1.tis) obyvatel, do 15 min. dojezd	Bernartice nad Odrou	I.	0,3848	4	1,5009	1,2358	2,0000	0,2935	Nový Jičín	1	Moravskoslezský kraj	1
Si5 (500, 1.tis) obyvatel, do 15 min. dojezd	Benešov u Semil	I.	0,2935	5	1,5009	1,2358	2,0000	0,2935	Semily	1	Liberecký kraj	1
Si12 (5.tis, 10.tis) obyvatel, nad 15 min. dojezd	Bechyně	II.	2,0000	1	2,0000	1,5120	2,0000	0,5970	Tábor	1	Jihočeský kraj	1
Si12 (5.tis, 10.tis) obyvatel, nad 15 min. dojezd	Doksy	II.	2,0000	1	2,0000	1,5120	2,0000	0,5970	Česká Lípa	1	Liberecký kraj	1
Si12 (5.tis, 10.tis) obyvatel, nad 15 min. dojezd	Jesenice	II.	2,0000	1	2,0000	1,5120	2,0000	0,5970	Černošice	1	Středočeský kraj	1
Si12 (5.tis, 10.tis) obyvatel, nad 15 min. dojezd	Kojetín	II.	0,9632	4	2,0000	1,5120	2,0000	0,5970	Přerov	1	Olomoucký kraj	1
Si12 (5.tis, 10.tis) obyvatel, nad 15 min. dojezd	Hostivice	II.	0,5970	5	2,0000	1,5120	2,0000	0,5970	Černošice	2	Středočeský kraj	2

Fig. 3.2 Statistical indicators of efficiencies

information about municipalities and he or she can compare of efficiency of selected municipalities.

The application can be used by three types of users:

(a) An anonymous user, without the need to register or provide any personal information
(b) A registered user with the role of analyst
(c) A registered user with the role of administrator

An anonymous user can either log in or select the option to use the functionality of calculating the efficiency comparison with other municipalities. For municipalities with which they compare the fictitious municipality with the selected parameters, the user sees only the calculated number, which is used for comparison, not the values of the parameters that were used for the calculation.

A registered user with the role of analyst can see the list of municipalities that contains basic information about each of them. This user can see all information about previous calculations and the value of all parameters used to calculate the municipalities' efficiency.

A registered user with the role of administrator has the same rights as the user with the role of an analyst. His or her higher rights allow him or her to edit users and registration, and change and delete data.

5 Conclusion

Public services are provided by a very heterogeneous spectrum of municipalities, represented by small villages with several dozens of citizens up to big cities. In order to make comparisons in as homogeneous groups as possible, categories of municipalities were proposed that reflect the size of the domestic population and the position of municipalities in the settlement structure.

The efficiency of libraries regardless of the categorization by municipalities reaches an average of 0.4458. In the case of the smallest municipalities, the mean efficiency values range from 0.2 to 0.8, which represents a significant inefficiency. It can be argued that the operation of most municipal libraries is ineffective with respect to the input and output data analyzed. As the size of the settlement increases, efficiency increases on average. The most effective are library services in towns and cities.

The segments with more inhabitants had higher efficiency of elementary schools in general. Only municipalities with inhabitants from 500 to 999 had the lowest efficiency, both with the driving distance below and above 15 minutes to the closest bigger town.

The findings are important for the discussion on how to set the availability of public services. At present, efficiency does not play a key role in capacity building. The establishment of a municipal library actually depends solely on an independent decision of the municipal council. The funding is conditioned by the willingness to cover the deficit of its operation with regard to the overall size of the municipal budget.

The operating costs of municipal libraries are covered often at the expense of other agendas performed by municipalities (e.g., in the form of sharing staff capacity). This means that some of the costs are not reflected in the financial statements associated with the library service, or are not even monetized. This means that the efficiency of small municipal libraries may in fact be even lower on average.

When it comes to the provision of school education, citizens of a community usually consider this role as a basic function of the municipality. It follows that even small communities, if they already have a school established (either in its entirety or only the first grade), strive to maintain them even at the cost of massive subsidies from their own budget. It is often the case that various exceptions apply, such as small classes, or teaching two years in just one year in order to maintain schooling. In the case of some rapidly developing municipalities close to large centers (affected by the process of suburbanization), the lagging of capacities behind the current need is observed. Therefore, the capacity of the core city is often used. If the municipality is in debt due to the construction boom, this behavior may be classified as a form of parasitism. The system of shared taxes does take into account the fact that the municipality operates a school when redistributing tax income among municipalities. However, this bonus covers only a fraction of the cost of running a school and is not an incentive to set up new or expand the capacity of existing schools. Some municipalities continue to enforce compensation payments for pupils commuting from other municipalities although this is not entirely in line with the law.

Through the administrative activities of municipalities of the third type (ORP—municipalities with extended competence), school districts are created, which define the commute area of individual schools. The aspect is the size of capacities and transport accessibility. However, the efficiency of schools does not play a crucial role in this decision-making.

The Ministry of the Interior, as the guarantor of the project, has initiated the introduction of various tools for monitoring and evaluating the efficiency of public administration activities. The created application represents a progressive tool for the application of digital technologies in public administration; it enables further evolution of this tool with regard to data availability.

Representatives of municipalities, as well as civil servants of the Ministry of the Interior, will obtain a tool enabling them not only to compare the efficiency of individual municipalities, but thanks to the extensive database and a map display also act as a tool for modeling of catchment circuits and service efficiency.

Acknowledgment This work was developed within the project "*Application of nonparametric methods (DEA) to analyze and to compare the efficiency of municipalities*" that is supported by TAČR (Technology Agency of the Czech Republic), Program ÉTA (project code TL01000463).

References

1. Hallerod B, Rothstein B, Nandy S, Daoud A (2013) Bad governance and poor children: a comparative analysis of government efficiency and severe child deprivation in 68 low- and middle-income countries. World Develop 48:19–31. https://doi.org/10.1016/j.worlddev.2013.03.007
2. Hladík M (2019) Universal efficiency scores in data envelopment analysis based on a Robust approach. Expert Syst Appl 122: 242–252. https://doi.org/10.1016/j.eswa.2019.01.019
3. Holmberg S, Rothstein B (2012) Good government: the relevance of political science, Edward Elgar Publishing. ISBN 9780857934925
4. Charnes A, Cooper WW, Rhodes E (1978) Measuring the efficiency of decision making units. Eur J Operat Res 2(6):429–444
5. Jablonsky J (2016) Efficiency analysis in multi-period systems: An application to performance evaluation in Czech Higher Education. Cent Eur J Operat Res 24(2):283–296. https://doi.org/10.1007/s10100-015-0401-z
6. Jetmar M (2018) Socioekonomická diferenciace obcí. Studie pro diskusi o členění obcí vytvořený pro potřeby realizace projektu „Uplatnění neparametrických metod (DEA, FDH) k analýze a ke komparaci efektivnosti obcí TL01000463 – registrovaného u TAČR
7. Lison B, Reip N (Co-Author) (2016) Research for Cult Committee – The New Role of Public Libraries in Local Communities, European Union, ISBN: 978-92-823-9864-7
8. Ozcan YA, Khushalani J (2017) Assessing efficiency of public health and medical care provision in OECD countries after a decade of reform. Cent Eur J Operat Res 25(2):325–343. https://doi.org/10.1007/s10100-016-0440-0
9. Peková J, Jetmar M, Toth P (2019) Veřejný sektor, teorie a praxe v ČR. Wolters Kluwer, Praha, 783 s., vydání první. ISBN 978-80-7598-209-4
10. Strategic Framework for the Development of Public Administration of the Czech Republic, Ministry of the Interior of the Czech Republic, 2016. https://www.mvcr.cz/clanek/strategicky-ramec-rozvoje.aspx
11. Act No. 257/2001 Coll. Act on Libraries and Conditions for the Operation of Public Library and Information Services (Library Act) as amended
12. Act No. 561/2004 Coll. Act on Preschool, Basic, Secondary, Higher Vocational and Other Education (School Act) as amended
13. Hamernikova B et al (2020) On statistical methods based information system for decision support of municipalities, In: 4th EAI International Conference on Management of Manufacturing Systems. Springer International Publishing, Cham, pp 205–218. EAI/Springer Innovations in Communication and Computing. ISSN 2522-8595. ISBN 978-3-030-34271-5
14. Jerabek M, Kubat J, Fabera V (2020) Smart, smarter, and smartest city: the method to comparison of cities. In: 4th EAI international conference on management of manufacturing systems.

Springer, Cham, pp 33–41. EAI/Springer Innovations in Communication and Computing. ISSN 2522-8595. ISBN 978-3-030-34271-5
15. da Cruz NF, Marques RC (2014) Revisiting the determinants of local government performance. Omega 44(C):91–103
16. Drew J, Kortt M, Dollery B (2015a) What determines efficiency in local government? A DEA analysis of NSW local government. Econom Pap J Appl Econom Policy 34(4):234–256. https://doi.org/10.1111/1759-3441.12118
17. Sousa MCS, Araújo PLCP, Tannuri-Pianto ME (2012) Residual and technical tax efficiency scores for Brazilian municipalities: a two-stage approach. Estudos Econômicos (São Paulo) 42(1):43–74. https://doi.org/10.1590/S0101-41612012000100002
18. Brettenny W, Sharp G (2016) Efficiency evaluation of urban and rural municipal water service authorities in South Africa: a data envelopment analysis approach. Water SA 42(1):11–19. https://doi.org/10.4314/wsa.v42i1.02

Chapter 4
Mobile Application for Remote HR Management

Angelina Iakovets, Michal Balog, and Peter Lazorik

1 Introduction

Actual situation in the world has impact on every economic field. Many companies were forced to introduce a home office, especially for administrative positions. The Government of the Slovak Republic took this issue especially seriously [1]. Stricter rules were accepted in February 2021—mandatory home office was provided for all employees to whom the work allows [1, 2]. The pre-rules and recommendations of the government were fixed in the Labor Code of the Slovak Republic. The Labor Code (No. 66/2020 Coll.) [3] introduces several changes in connection with the coronavirus and gives a clear definition of the concept of a home office, where home office is the employment relationship of an employee who performs work for the employer according to the conditions agreed in the employment contract, either at home or at another agreed place [2]. The chapter No. 66/2020 defines some types of homework for employees: domestic work or telework is work performed regularly from the employee's household. It is not extraordinary or occasional work performed by an employee, but usual work performed at home with the consent of the employer and in agreement with him or her. Such work from home will be called "home office" [4].

The dividing criterion between housework, telework, and home office is the regularity of work outside the workplace. For example, it can be determined by how often or how long on a regular basis each week an employee performs housework, domestic work, or telework. And the employment contract should respond to that [4].

A. Iakovets (✉) · M. Balog · P. Lazorik
Faculty of Manufacturing Technologies, Technical University of Kosice,
Presov, Slovak Republic
e-mail: angelina.iakovets@tuke.sk; michal.balog@tuke.sk; peter.lazorik@tuke.sk

L. Knapcikova et al. (eds.), *Sustainable Management of Manufacturing Systems in Industry 4.0*, EAI/Springer Innovations in Communication and Computing,
https://doi.org/10.1007/978-3-030-90462-3_4

Many enterprises, especially manufacturing ones, are faced with the fact that they do not have sufficient personnel control resources. Personnel control had to be carried out remotely, since the epidemiological situation did not allow doing it "on the spot" as it was before. The current situation can be perceived critically, or the experience of other countries and enterprises from different economic spheres can be consolidated and an effective system of remote personnel control can be drawn up. The controlling role usually belongs to the HR manager or the director, depending on the size of the enterprise. Software for remote control and regulation of work processes could become a regulating tool labor relation.

1.1 Formulation of the Problem

Other duties of the employer are defined by the current Labor Code of Slovak Republic [4]:

- Maintenance of technical and software equipment, if the employee does not use his or her own—his or her own technical and software equipment carries risks.
- Ensuring the protection of processed data—applies not only to personal data, but also to all data processed by employees.
- Informing the employee about the restrictions related to the use of equipment and software.

All Internet articles and comments [1–4] to the law are about way of defining labor relations and rules of home office. The government does not give an answer, and all comes down to the fact that the employer and worker should agree on the rules of working conditions. This phenomenon is explained by the fact that earlier in Slovakia and other European countries, remote work was not so widespread [5] (see Fig. 4.1).

Work from home in the EU in 2017 showed a positive dynamic, but over time this process has stalled slightly [5–7]. Statistic data of 2017 shows that about 17% of people wanted to work remotely [5], but according to Fig. 4.1. Slovak Republic (green scale on Fig. 4.1.) represented only slightly more than 3% of remotely working people. The low percentage of employed remotely is due to the lack of software base and the specific of business style introduction in this area.

Since 2017, the proportion of those working from home has increased, and the age of these people has also increased. In the EU, only 1.8% of young people aged 15–24 in 2018 usually worked from home, compared with 5.0% among 25–49 and 6.4% among 50–64 [5, 6].

The pandemic has changed the dynamics of the spread of the idea of remote work in various professions. As mentioned above, the legislative base as well as the general epidemiological situation in the world has become a decisive factor [6, 7].

Telework has become a tool, one which is able to stop a work pattern interruption of all enterprises, caused by the epidemical situation of 2019–2021 [7].

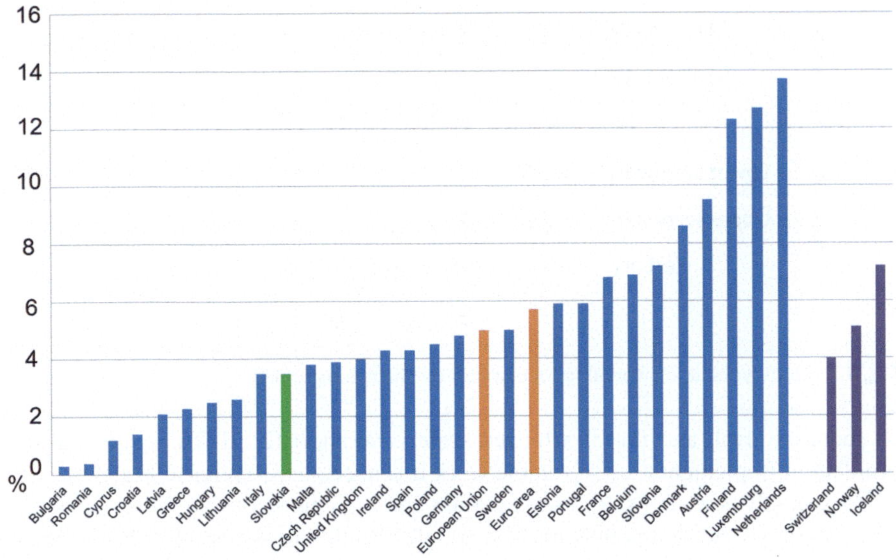

Fig. 4.1 Employed persons aged 15–64 usually working from home in the EU, 2017 [5]

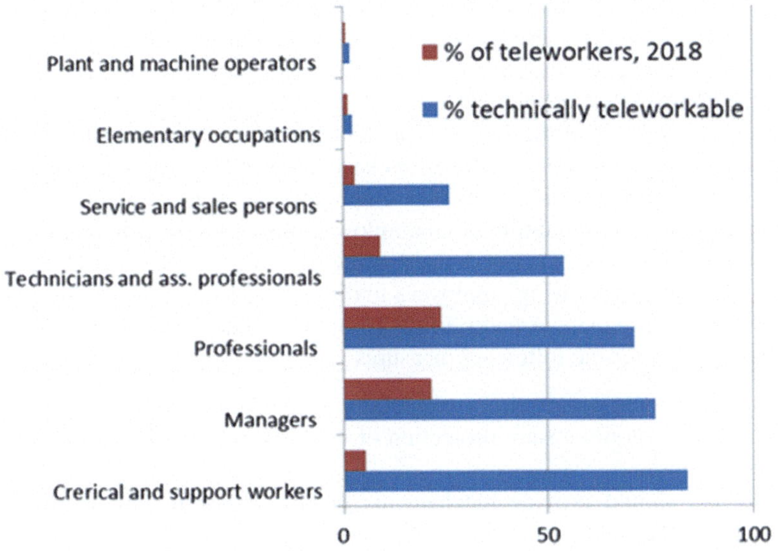

Fig. 4.2 Prevalence of telework by occupation, 2018 [8]

The studied area is Slovak Republic, and it is well known by manufacturing enterprises. Fig. 4.2. represents the ability of providing remote work by occupation. Despite the fact that major workers are teleworkable, the technical base is still not

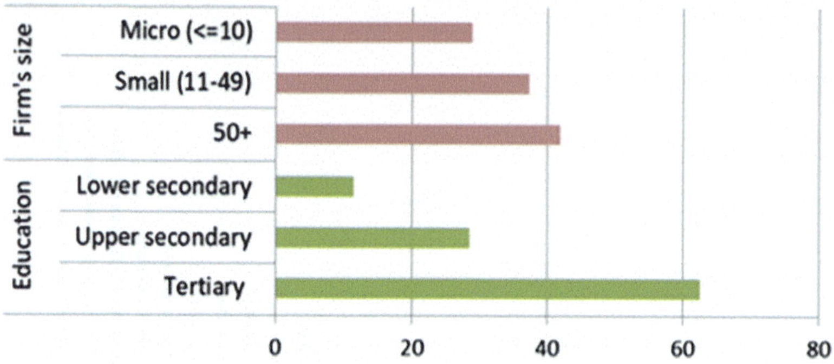

Fig. 4.3 Employees in teleworkable occupations by characteristics [8]

on required level, and that is why red scale is so short [2, 3]. To understand the problem of implementation of remote work it is necessary to refer to Fig. 4.3.

As specialists write [8, 9], only medium and big enterprises are most teleworkable and only educated people are able to adapt to this work style [8]. With regard to the percentage of using telework by sector, manufacturing is in the lowest position [7]. It can be explained by the fact that manufacturing companies, with spreading latest technologies, are accustomed to fieldwork. Enterprises try to be flexible to customers' needs, and it requires the presence of employees at the workplace.

37% of EU workers started working from home during isolation [10]. During this period, a number of discontents arose both on the part of employees and on the part of the company's management.

Managers tried to control the work process and often this brought a number of inconveniences and the working day often lasted longer. Enterprise managers asked themselves the question of control over employees, as well as the duration of the working day and the availability of an employee outside working hours.

The right to disconnect is not defined in the EU law; that is why EU parliament plans to provide law allowing employees to disconnect from work during nonwork hours without consequences and setting minimum standards for remote work. EU parliament called for the following measures:

- Employers should not require workers to be available outside their working time.
- EU countries should ensure protection of workers from victimization and other repercussions.
- Employers should provide mechanisms in place to deal with complaints or breaches of the right to disconnect.
- Remote professional learning and training activities must be counted as work activity and must not take place during overtime or days off without adequate compensation [10].

In order to offer an effective solution for monitoring and supporting employees, it is worth referring to the experience of countries and enterprises where remote work was initially more common.

For example, the British company Founders Forum interviewed part of the population that worked remotely during isolation. Most of the respondents were asked: "How would you like to continue working?" Over 92% of respondents said that they would like to be able to work from home to some extent [11]. Athena Marousis writes in her article that statistics show that workers who initially worked remotely have higher productivity and easily combine personal life and work [12]. In turn, this means that the remote work mode can be effective if you choose the right software.

1.2 HR Controlling Methodology

Human sources in every enterprise take the main role. Valuable personnel provide sustainable income for the company and the quality of goods and services; therefore it is important to ensure not only control over the employee's work, but also his or her support and training.

The pandemic caught many enterprises by surprise and few people managed to train employees to use the software, and this slowed down the work process. Any remote work should be comfortable for employees at all levels.

Management control system in the HR management system is not just control over personnel and the implementation of planned indicators, but a management system that ensures the development of a clear system of plans and tasks, fixing these plans and tasks, monitoring their implementation, and taking advisory measures in case of deviation.

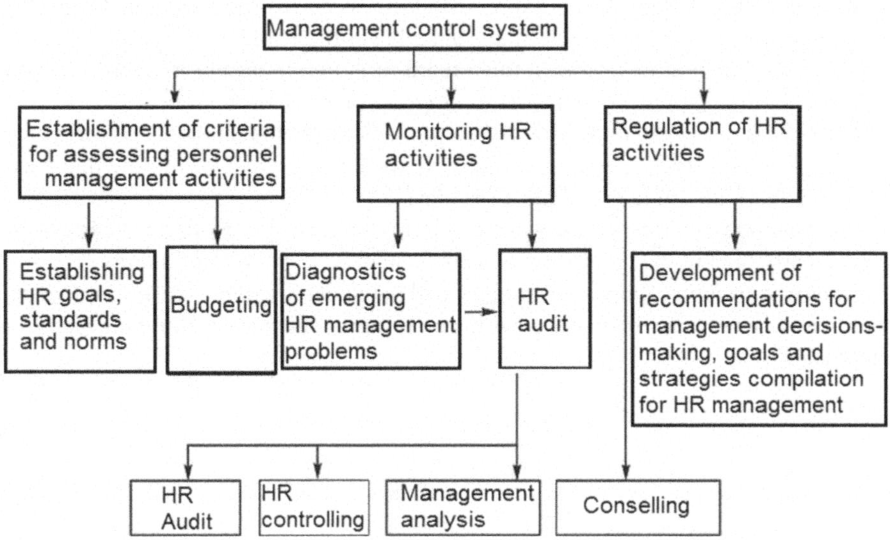

Fig. 4.4 Content of personnel controlling in the personnel management system [14]

According to Fig. 4.4., management control system manages organizational resources like human, physical, financial, and also organization as a whole. Due to the fact that all the components are interconnected, it is important not only to control personnel, but also to quickly respond on personnel needs and provide all-round support. All successful firms, where telework was provided many years ago, constantly support employee development. According to the latest study of Athena Marousis, the biggest part of respondents categorizes itself as "visual" (66%); it means that they better learn by visual materials as videos or pictures [12]. Uncertainty is what most people surveyed felt over the past year. Last two years, 65% of people believed that in the near future they can lose their jobs [12]. The main reasons for this opinion are uncertainty in their qualifications, fear of getting sick, feeling of pressure due to the current situation in the world, etc. In such a difficult period of time, it is important for entrepreneurs to protect employees from stress and keep the staff in full. In the nearest future would be provided understandable software for people of different ages and different levels of education it is worth considering that the graphic material should took the biggest part.

Main tasks should be solved:

- Time management
- Emphasis on mental health support
- Team check-ins
- Focus on performance management data
- Employee training [13]

Effective time management and scheduling will help employees evenly share the load and not lose productivity.

Emphasis on mental health support. The CDC found that more than 40% of adults in the United States suffer from mental illness during the pandemic [15]. This situation is not only in the United States but all over the world; every employee now feels a special mental pressure. In this regard, it is important not to overdo it with the control of the employee and do it so as not to provoke even more stress.

Team check-ins. Virtual meetings of employees will help maintain a positive attitude, as well as distribute responsibilities, as well as everyone will know that each employee faces the same problems every day.

Focusing on performance management data. Collecting performance data helps to efficiently plan in the enterprise as well as quickly respond to problems.

Employee training. Employee training will help not only improve the employee's qualifications but also help cope with emerging problems or eliminate gaps in knowledge.

1.3 Mobile Applications

By 2021, there is probably no such person who has never encountered a mobile application. The number of users of mobile devices and mobile applications is growing every year and the number of older users is also growing. Easy and intuitive interface engages users of all ages. The mobile application, by its nature, can be easily installed on various devices supported by the mobile OS [16].

Market research of market employee engagement software has shown that the top 10 applications are:

- KazoopHR
- 15Five
- Reflective
- Peakon
- OfficeVibe
- 7Geese
- GLINT
- Lattice
- Emplify
- Honestly [17]

All these programs have some features that are very helpful even for manufacturing enterprises as:

- Customizable features
- Easy to use
- Feedback
- Real-time scores [17]

Internet journal Best Money Moves writes that HR leaders should introduce initiatives in the next 12 months:

- Employee engagement
- Training and support for managers
- Diversity, equity, and inclusion (DE&I) programs
- Employee learning and development
- Performance management

Most of these initiatives revolve around adapting the procedures and processes established to limit the risk of COVID-19 while remaining operational [18].

HR managers also have a wide scale of mobile applications, which provide support of their daily duties, such as WorkBright, Namely, Zenefits, Halogen TalentSpace, Justworks, ClearCompany, Optimum HRIS, Gusto, Deputy, WorkStyle, Cezanne HR, APS OnLine, Kronos Workforce Ready Suite, Fairsail, SAP SuccessFactors, iAppreciate, Zuman, Engagedly, and i-Sight [19].

The main problem of all mobile applications mentioned above is that they are general, do not cover the needs of each enterprise, and are also targeted at a certain level of the employee. The main disadvantage of a mobile application is that one application cannot replace a full-fledged enterprise management program at various levels, especially for manufacturing enterprises.

The use of a large number of applications overnight brings a number of problems: the user must be erudite in this area, easily perceive new information, and be an active user of mobile devices. This level of erudition is very difficult to achieve, especially due to the specifics of the work of some employees.

2 Mobile Application for Sustainable Remote HR Controlling Process

Since the study was aimed at the manufacturing sector, an important criterion was to make an intuitive interface and not burden the employee with multitasking applications. The surveyed enterprises planned to provide supervision and support for assembly shop employees and welders who could not work remotely. The volume of manufactured products and losses associated with product quality did not allow to exclude personnel control.

Previously, the company carried out automated attendance control. Upon arrival at work, employees put cards to the scanner and thereby recorded the beginning of the working day, the end of the working day, breaks, etc. The best solution was to use the existing accounting system for the mobile application. Creating a new database of passwords and accounts would take a long time and entail a number of log-in problems. Employee cards were equipped with an NFC chip, so registration in the application was also carried out using these cards. This solution reduced the time of user identification in the system [20–22].

This method of logging into the application is especially convenient for employees using security tools (Fig. 4.5).

NFC technology is widely used in banking applications for making contactless payments. This type of application allows users not to use payment cards and smartphones for the same purpose. Thus, the client instantly receives information about payments. Applications of this type work using an Internet connection in order to update the information to which the card is linked. Other NSF card reader applications do not require the Internet, but the information they give about the card is narrowly limited [23].

Card opens the access to HR manager profile or operator profile. Depending on the access rights, the user menu with its own subcategories is opened (see Fig. 4.6).

Figure 4.6 illustrates the menu available to the operator and manager. The operator has accessed only performing tests, video instructions, and work shift management.

Managers accessed edition and creation tests and instructions, statistics of the operator's work shift, and reporting of each employee.

Fig. 4.5 Log-in within NFC employee card

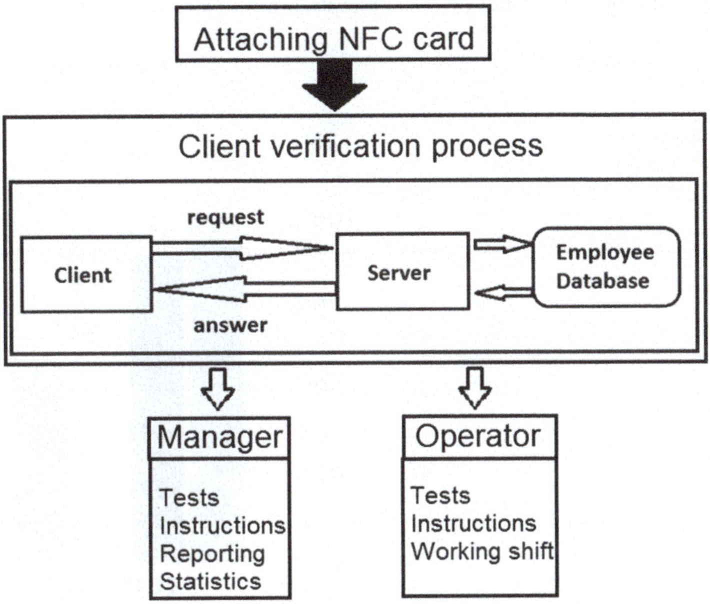

Fig. 4.6 The process of opening access with a card

As soon as NFC card opens the access the working shift of each employee is automatically started and data of surfing by application is collected to the cloud of the enterprise. The "working shift" menu of the operator has several options, as:

- Check-in
- Registration of breaks
- Registration of processed product number
- Ending the work shift

This data collection approach did not burden employees with undue direct managerial oversight.

Mobile application also collects the test results and information about employee productivity. The data collection speed directly depended on the speed of the Internet connection, since for the convenience of use the application was of a hybrid type. Such an application works using the Internet connection and instantly transmits data, which allows you to always have real information at hand. Also, the hybrid type of application allowed part of the data to be saved off-line to the device in case of problems with the Internet connection [16].

It was assumed that the proposed application would be convenient to use, since it fulfilled several criteria: a large amount of graphic material, a minimum of manual text input, and an easy way to enter the application. Also, the statistical data in Fig. 4.7 show that the number of users of mobile applications is very diverse (from 18 to 65+ years old), which falls under the age of the active working population. Managers were more adapted to professional software, so more emphasis was placed on operators, since their number is predominant in manufacturing enterprises.

To assess the quality of the proposed application, users were asked to complete the form.

Fig. 4.7 Smartphone users by age structure [23]

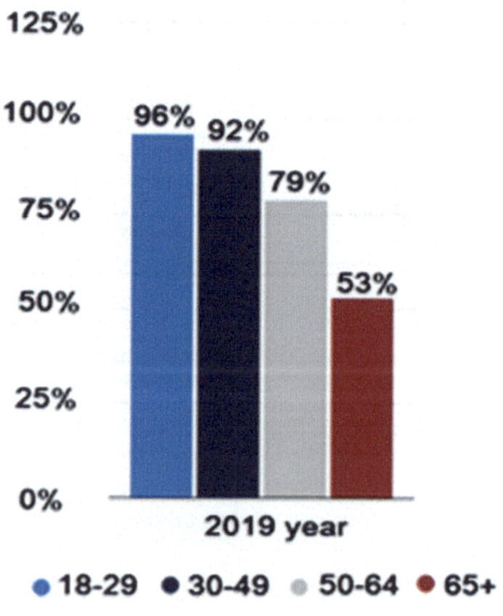

Form of System Usability Scale (SUS) is a world-known form of getting feedback from mobile application users [24–26]. SUS consists of 10 items with five options for respondents (from strongly agree to strongly disagree). This method was originally created by John Brook in 1986. The SUS evaluates a wide range of products and services, including hardware, software, mobile devices, websites, and applications [26]. The user's score for each question is converted to a new number, added, and then multiplied by 2.5 to convert the original scores from 0–40 to 0–100. Although the scores are 0–100, they are not percentages and should only be considered in terms of their percentile ranking. According to research, an SUS score above 68 is considered above average, and anything below 68 is considered below average.

The average score of 50 SUS tests was 28 points. After converting this value by multiplying by 2.5, 70 was obtained, where 70 is more than 68, which means that the result is above average and is satisfactory. The next step was the illustration of conversion results (see Fig. 4.8).

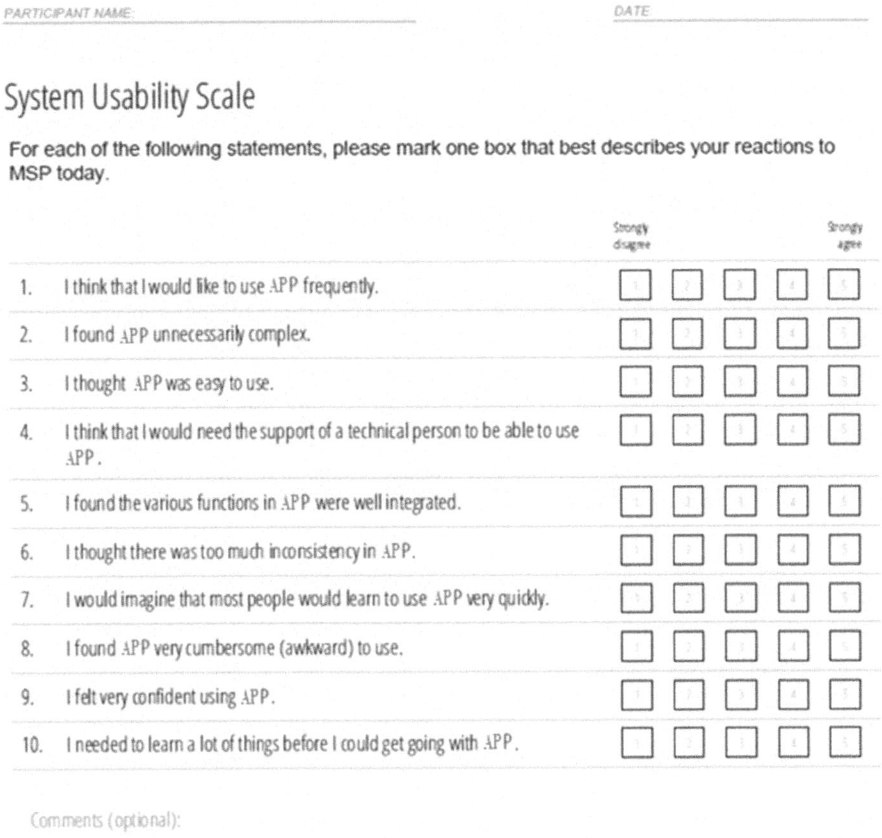

PARTICIPANT NAME: DATE

System Usability Scale

For each of the following statements, please mark one box that best describes your reactions to MSP today.

		Strongly disagree				Strongly agree
1.	I think that I would like to use APP frequently.	1	2	3	4	5
2.	I found APP unnecessarily complex.	1	2	3	4	5
3.	I thought APP was easy to use.	1	2	3	4	5
4.	I think that I would need the support of a technical person to be able to use APP.	1	2	3	4	5
5.	I found the various functions in APP were well integrated.	1	2	3	4	5
6.	I thought there was too much inconsistency in APP.	1	2	3	4	5
7.	I would imagine that most people would learn to use APP very quickly.	1	2	3	4	5
8.	I found APP very cumbersome (awkward) to use.	1	2	3	4	5
9.	I felt very confident using APP.	1	2	3	4	5
10.	I needed to learn a lot of things before I could get going with APP.	1	2	3	4	5

Comments (optional):

Fig. 4.8 Template of SUS test of application

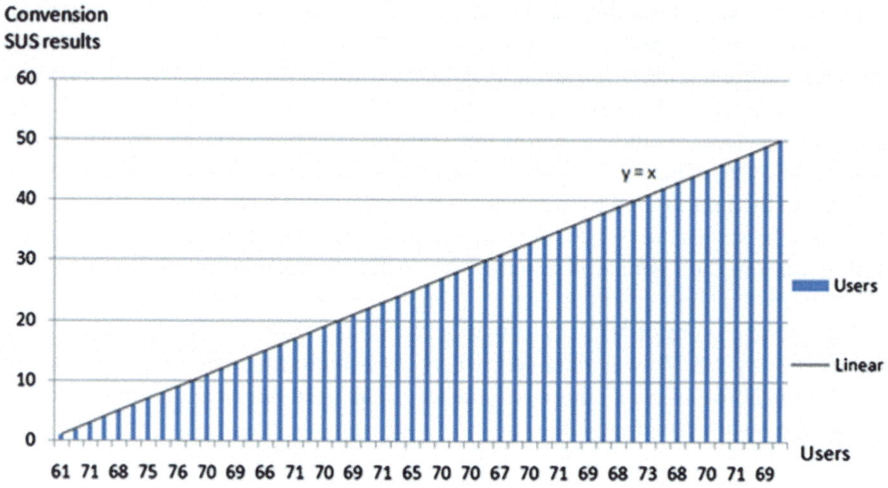

Fig. 4.9 Conversion of SUS test

The data from graph (Fig. 4.9) had shown a positive trend, which means that in the future an increase can be expected in user satisfaction with the application.

3 Conclusions

Cloud technology is replacing legacy systems, and major vendors plan to start new investments in relationship management tools. The proposed application can serve as an additional tool for monitoring and supporting employees, but it cannot be the only universal solution. Statistics and literature review have shown that many companies use a combination of different applications and this is effective. The proposed application is suitable for manufacturing enterprises, where operators are involved in the production process, and managers, due to a pandemic situation, cannot quickly respond to all changes in production processes. The application is also useful for managers after the pandemic, as it helps operators to have training materials always at hand and also reduces the number of paper reports issued each shift. Remote collection of information by the application will help to have a complete picture of HR sources of the enterprise, as well as collect the necessary information without stopping the production process.

The peculiarity of the mobile application is in its updating versions, where each update can introduce new functionality as well as adapt the application for various carriers (mobile phones, VR and AR glasses, tablets, and other mobile devices).

In production, where the cost of operator failure carries huge costs, it is very important to be able to quickly respond to changes in the quality of personnel, technology, etc. Mobile devices can be used not only as a container for an application, but also for other purposes as data collection from sensors and machines, for scanning the information, and so on. By its nature, any mobile device makes the employee less attached to the workplace and thus helps to do their job remotely and quickly respond to the problem at the point of its occurrence.

The use of mobile applications is another step towards the diffusion of the latest Industry 4.0 technologies.

References

1. TA3, Home office is mandatory from Monday, companies will check, 2021., https://www.ta3.com/clanok/1202530/home-office-je-od-pondelka-povinny-firmy-budu-kontrollovat.html, last accessed 2021/02/10
2. Pohorela P: Work from home (so-called home office) – what rules apply to it. Podnikajte.sk. https://www.podnikajte.sk/pracovne-pravo-bozp/praca-z-domu-home-office-pravidla, last accessed 2021/02/10
3. Bartko T: Amendment to the Labor Code adopted on 4 April 2020 in connection with the coronavirus. Podnikajte.sk. https://www.podnikajte.sk/pracovne-pravo-bozp/novela-zakonnika-prace-4-4-2020-koronavirus, last accessed 2021/02/10
4. Securion; News –home work and telework. https://www.securion.sk/zmeny-zakonnik-prace-2021-gdpr-a-ine/, last accessed 2021/02/10
5. Eurostat: Your key to European statistics, Working from home in the EU, 2017. https://ec.europa.eu/eurostat/web/products-eurostat-news/-/DDN-20180620-1, last accessed 2021/02/10
6. Eurostat: How usual is it to work from home 2018. https://ec.europa.eu/eurostat/web/products-eurostat-news/-/DDN-20200206-1, last accessed 2021/02/10
7. European Comission: Telework in the EU before and after the COVID-19:where we were, where we head to. Science for policy briefs. https://ec.europa.eu/jrc/sites/jrcsh/files/jrc120945_policy_brief_-_covid_and_telework_final.pdf, last accessed 2021/02/12
8. Milasi S, Bisello M, Hurley J, Sostero M, Fernández-Macías E: The potential for teleworking in Europe and the risk of a new digital divide. 2020. Voxeu.org. https://voxeu.org/article/potential-teleworking-europe-and-risk-new-digital-divide, last accessed 2021/02/13
9. OECD Policy Responses to Coronavirus (COVID-19). Productivity gains from teleworking in the post COVID-19 era: How can public policies make it happen. http://www.oecd.org/coronavirus/, last accessed 2021/02/13
10. European Parliament: Parliament wants to ensure the right to disconnect from work. 2021. https://www.europarl.europa.eu/news/en/headlines/society/ , last accessed 2021/02/13
11. Wood T: The Future of Remote Work, According to Startups. Technology. 2020. https://www.visualcapitalist.com/the-future-of-remote-work-according-to-startups/, last accessed 2021/02/11
12. Marousis A: Would you take a pay cut to keep working remotely. 2021. Talentlms. https://www.talentlms.com/blog/remote-work-statistics-survey/, last accessed 2021/02/11
13. Merchant AK (2017) In: Van der Stede WA (ed) Management control systems: performance measurement, evaluation and incentives, 4th edn, Harlow, England. ISBN 9781292110554. OCLC 965154191
14. Merchant AK: Management control systems: performance measurement, evaluation and incentives. Van der Stede, Wim A. (Fourth ed.). Harlow, England. 2017. ISBN 9781292110554. OCLC 965154191
15. HR Trends: The Biggest HR Challenges for 2021. Primalogic. https://primalogik.com/blog/the-biggest-hr-challenges-for-2021/ , last accessed 2021/02/11
16. Blokdyk G: Mobile Operating System A Complete Guide. 2020. 241 p. ISBN: 9781867367420
17. OECD Data. Enterprises by business size. 2017. Available online: https://data.oecd.org/entrepreneur/enterprises-by-business-size.htm, last accessed 2021/02/11

18. Best Money moves: Top HR Challenges in 2021 and How to Overcome Them. 2021. https://bestmoneymoves.com/blog/2020/10/20/top-hr-challenges-in-2021-and-how-to-overcome-them/, last accessed 2021/02/08
19. Hanks C: The 52 Best Mobile Apps for Human Resource (HR) Managers. Wonolo. https://www.wonolo.com/, last accessed 2021/02/08
20. Husár J, Hrehová S, Kaščák P (2020) Design of Concept for transport and monitoring of biological samples using RFID technology. New approaches in Management of Smart Manufacturing Systems. Springer, Cham, pp 91–111. https://doi.org/10.1007/978-3-030-40176-4_6. IPrint ISBN 978-3-030-40175-7. Online ISBN 978-3-030-40176-4
21. Židek K, Piteľ J, Adámek M, Lazorík P, Hošovský A (2020) Digital twin of experimental smart manufacturing. Assembly System for Industry 4.0 Concept. Sustainability 12(9):3658. Industry 4.0 for SMEs – Smart Manufacturing and Logistics for SMEs. https://doi.org/10.3390/su12093658
22. Seasia: 7 Best NFC Payment Apps That Provides an Extra Layer of Security. https://www.seasiainfotech.com/blog/best-nfc-payment-apps/, last accessed 2021/02/08
23. Gridasov A: Personnel assessment methods and criteria: how to assess the performance of employees in an organization. 2020. Calltouch blog. Available online: https://blog.calltouch.ru/otsenka-personala-metody-kriterii-otsenki-raboty-personala-v-organizatsii/, last accessed 2021/02/08
24. Mifsud J: Usability testing of mobile applications: a step-by-step. Usability Geek. Available online: https://usabilitygeek.com/usability-testing-mobile-applications/ (last accessed on 10 September 2020)
25. TestMatick. Usability-testing mobile application. 2018. TestMatick. Available online: https://testmatick.com/ (last accessed on 10 September 2020)
26. Jacob Nielsen J, Laranger H: Web Design: Usability of Web Sites = Prioritizing Web Usability. - M.: "Williams", 2007. - S. 368. - ISBN 0-321-35031-6

Chapter 5
Mathematical Modeling of Drive and Dynamic Load of Teeth of Cylindrical Worm Gear

Slavko Pavlenko, Jozef Mascenik, and Tibor Krenicky

1 Introduction

Worm reducers are used in the case of drives of a number of modern machinery of diverse technological direction, especially in mechanical engineering, construction industry, road transportation, agriculture, metallurgical engineering, chemical and food industry, consumer industry, etc. Their technical level and load capacity in a high degree determine technical, economical, and operational characteristics of machines, the part of which they become after assembling. Thus the need to assure increase of service life of worm reducers represents a significant and up-to-date task [1].

Modern scientific researches in the field of design proposal and structure of machines move in direction of the development of methods of phenomena analysis. The development is connected with development of the branches as follows: general mechanics, mathematics, theory of optimization, mechanics of continuum, tribology, theory of stochastic processes, and theory of reliability, experimental methods, and methods of research of particular machines under actual operating conditions. The conditions can be either actual or created artificially or are mathematically modeled, i.e., realized by means of so-called simulation research with the application of modern computer technology.

Dynamics of machines perceived in a modern manner represents dynamics of systems requiring a dual approach: on the one hand, it is a detailed research of individual parts of the system, and on the other hand, it is a complex analytical research of the entire system.

S. Pavlenko · J. Mascenik (✉) · T. Krenicky
Faculty of Manufacturing Technologies with a seat in Presov, Technical University of Kosice, Presov, Slovakia
e-mail: slavko.pavlenko@tuke.sk; jozef.mascenik@tuke.sk; tibor.krenicky@tuke.sk

L. Knapcikova et al. (eds.), *Sustainable Management of Manufacturing Systems in Industry 4.0*, EAI/Springer Innovations in Communication and Computing,
https://doi.org/10.1007/978-3-030-90462-3_5

Modal analysis is a modern branch of dynamics [2] which for description of oscillatory processes and of oscillating behavior of engineering structures and of their parts uses the possibility of disintegration of a complex oscillating process into partial, i.e., modal, constituent. Each constituent is consequently characterized by modal frequency and by modal shape of oscillation.

2 General Principles: Worm Gears

Worm gears are designed to transfer torsional moment between skew shafts mostly at right angle. The worm gear which is located in a solid locked casing represents a worm gearbox (Fig. 5.1). The worm gearbox can consist of a single or several worm gears. Except for worm gears it may contain other toothed gears as well. The solid casing represents a support for capturing actions of forces from positioning of shafts which carry the segments of worm gear in precise axial directions, protects the gear against impurities, and generates conditions for its effective lubrication [3].

From a structural point of view, the worm gears are produced as independent nodes which are fixed to a joint frame with driving machines or other driven machines or as an integrated structure with a drive or other parts of machines in a single unit. Flange method of fixation can also be observed [4].

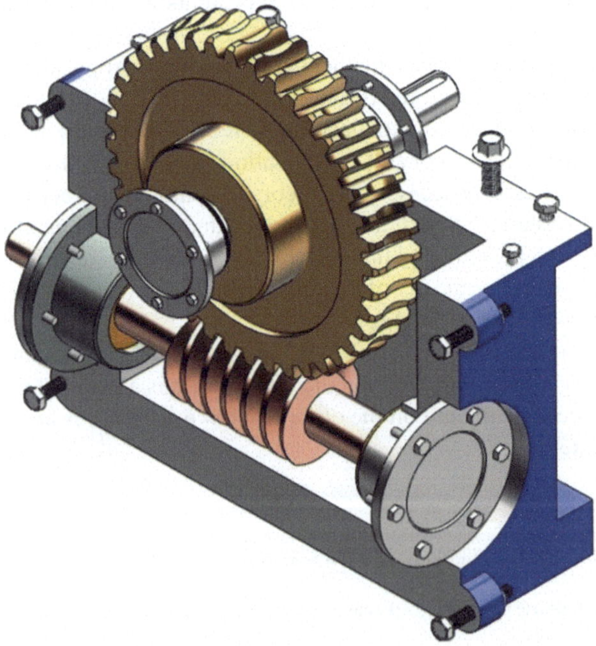

Fig. 5.1 Worm gears

Worm gears designed for general utilization are produced with particular parameters in accordance with norms and standards referring to application in diverse machineries even of special determination in case of which specific requirements are imposed upon their operation and working modes.

Theoretically, all of these worm gears can have a random angle between axes yet the most broadly utilized are the gears with mutually perpendicular axes.

Continuity of running is reached through good adaptive ability of a pair of segments of worm gear. One of the main advantages of worm gear is the possibility of self-locking in case of change of direction of transfer of performance by means of gear which is rather significant in a number of machines, especially in the case of lifting device. Therefore, the gears of some of the devices contain self-locking worm gear although its use could be avoided without any losses [5].

The main disadvantage of worm gears rests in the inevitability of utilization of costly and deficient materials for worm wheels (preferably on the basis of copper and tin) and low efficiency due to high friction losses in toothing which might amount to 70%. In this case high amount of heat is produced which causes fast heating up of the gearbox. This frequently results in the need of forced cooling of the gearbox, in extension of its size despite the fact that from the point of view of hardness it is not necessary.

In a high degree, friction coefficient depends on slip speed and on conditions under which oil layer (cotter) is formed in toothing.

Increase of slip speed causes a sharp drop in friction coefficient, which consequently results in increase of the gear efficiency. On the basis of the aforementioned details, it is obvious that worm gears are used in high-speed shafts. The selected values of friction coefficient between steel worm and worm wheel made of lead bronze for diverse slip speed values are given in Table 5.1.

Mean values of efficiency of concurrent worm gears (including losses in shafts caused by friction):

Number of runnings of a gear	Efficiency
$z_1 = 1$	$\eta = 0.7 \div 0.75$
$z_1 = 2$	$\eta = 0.75 \div 0.82$
$z_1 = 3 \div 4$	$\eta = 0.82 \div 0.92$

Table 5.1 The selected values of friction coefficient

Slip speed v_{sk} [$m.s^{-1}$]	Friction coefficient f	Slip speed v_{sk} [$m.s^{-1}$]	Friction coefficient f
0.01	0.01–0.12	2.5	0.03–0.04
0.1	0.08–0.09	3	0.028–0.035
0.25	0.05–0.0075	4	0.023–0.03
0.5	0.055–0.065	7	0.018–0.026
1	0.045–0.055	10	0.016–0.024
1.5	0.04–0.05	15	0.014–0.02
2	0.035–0.045		

In case of low performance, the decline in efficiency is not of significant meaning, yet in case of mean performances, the use of worm gears becomes less economically advantageous contrary to the use of toothed gears with spur toothing and thus the performance of worm gears does not exceed 50 kW and scarcely ever reaches 100 ÷ 150 kW.

2.1 Geometry of Worm Gears with Cylindrical Worm

Currently the worm gears with cylindrical or with globoid worm are the most frequently used.

Transfer ratio by application of a single degree of worm gear can commonly range within 8 ÷ 100; rarely it can reach the value of even 1000 which requires compact reconstruction of the gearbox.

In the production of worm wheel by a hob the profile of which is identical with one of the worms, the gear racks form diverse profiles of teeth in corresponding cuttings, and in cutting corresponding to the gear rack with the straight-line profile an evolvent worm is formed [6].

Each gear rack corresponds to its meshing line and in the case of a gear rack with straight-line profile it is a straight line. The total of meshing lines of all gear racks forms a meshing surface of the worm gear. The idea of a mesh of a pair of worm wheels as of a continuous set of gear racks leads to the assumption that the contact of working surfaces of worm runnings and of teeth of worm wheel occurs along continual skew curves which are referred to as tangent lines. Each tangent line is a geometrical area of points of the meshing surface in which tangency occurs between teeth of the worm wheel and runnings of the worm at the particular moment. Likewise, each position of the worm corresponds to the position of its tangent line which is part of the meshing surface [7].

Central plane Q (Fig. 5.2) divides meshing surface into two parts, i.e., the input one and the output one. The input part is the one where the threads (runnings) of the worm enter the meshing with the teeth of the worm wheel. Achieving either of the screw surface type depends in the case of the worm thread on the means of production which is selected according to the purpose of the gear use, on the volume of production, and on the means of finishing of the surface of threads. In the case of power gears, the most suitable is the use of evolvent worms with ground surface.

3 Dynamics of Worm Gears

Despite the fact that worm gears are extensively applied in practice, the professional literature only rarely offers publications related to the research of the gears from the point of view of their dynamics. The exception to the rule is the work by V. Klima [8], in which the author presents the necessity of taking into consideration the

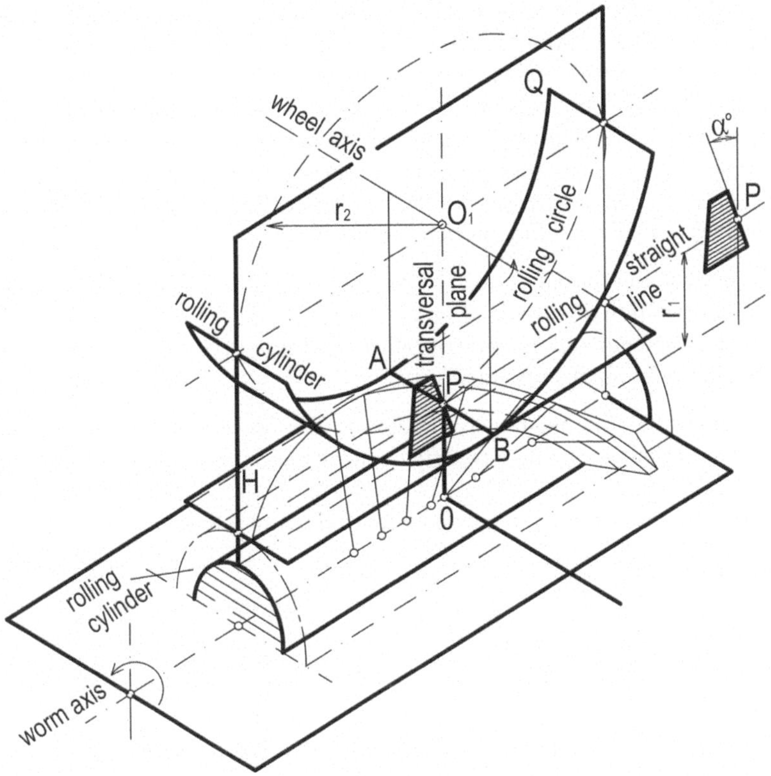

Fig. 5.2 Geometry of worm gears with cylindrical worm

influence of dynamic effects of teeth meshing on the determination of maximal values of loading forces.

For the case of teeth meshing of cylindrical worm gear system the more positive influence of dynamic effects during the start of teeth meshing is presented in [8]. In the case of spur-toothed wheels the circumferential speed of both meshing teeth is of the identical value and of identical direction in a pitch point. In the case of mesh of teeth of cylindrical worm gear systems the relative speed of faces of teeth of worm is at an angle of 90°- γ (with γ standing for lead angle of worm on pitch cylinder) towards the direction of circumferential speed of worm wheel at pitch diameter. Except for the aforementioned, even the circumferential speed of worm is several times higher than the circumferential speed of worm wheel and therefore the onset of faces of teeth of worm onto wheel teeth is more continuous and lacks negative dynamic effects contrary to spur wheels. The author states that at larger lead angles of worm or in case of higher rotations of worm it is inevitable to take into consideration dynamic effects as well. The author recommends to apply the method of replacement of mesh of cylindrical worm gear system by mesh of a gear rack with worm wheel and to use derived relations for determination of dynamic effects in case of spur gear systems [9].

Out of the foreign authors dealing with dynamics of worm gear A.W. Tuplin and N.L. Vejc should be mentioned. The other of the two elaborated in his publications is mainly the dynamics of self-locking worm gears [10–12].

As it has already been mentioned, the literature [13] offers the method of calculation of magnitude of additional dynamic force in case of worm toothing. To determine the yielding property of worm gear Tuplin applies the relation to determine the yielding property of teeth of spur-toothed wheels with skew teeth. The author takes into consideration the yielding property as follows:

- From deflection of tooth as of fixed beam
- From deformation of surface of faces of teeth
- From shift of teeth with regard to bodies of wheels
- From circumferential stress in a rim

Apart from the aforementioned yielding properties, the author also offers relation for calculation of yielding property occurring as a result of worm flexure due to loading.

Additional dynamic force is given by the following relation:

$$F_{dj} = \frac{A_b e_b}{G_z} + \left(F_j \text{ or } \frac{A_s e_s}{G_z}; \text{ lower of the values} \right) \tag{5.1}$$

with

e_b—total of permitted deviations of pitches of adjoining teeth [m]
e_s—periodical error of gear (profile deviation) [m]
G_z—yielding property of worm gear system in direction of the tangent towards the pitch circle of worm wheel per unit of length [m^2N^{-1}]
A_b, A_s—diagram values ([14] diag. 10)
F_j—circumferential force of worm wheel per unit of length [Nm^{-1}]

The [11] offers detailed analysis of the follow-up of self-locking worm gears. The follow-up is characterized as motion with declining kinetic energy and machine halt. From the point of view of direction of transfer the follow-up of self-locking gear can be divided into two types as follows:

1. The follow-up with transfer direction of torsional moment identical with the one in case of steady running conditions
2. The follow-up with the change of transfer direction of torsional moment

A lifting mechanism with self-locking worm gear which drives the winding drum can serve as an example. Standard mode refers to load lifting and unbraking mode is connected with lowering of a load. In the first case, the worm is a driving element and the worm wheel is the driven one. In the second case, the worm as well as the wheel represent the driving elements. Special emphasis is put on the determination of conditions related to unbraking. The study [12] presents conditions under which the unbraking mode occurs. The mode is typical for a sharp increase of accelerations or decelerations which practically results in instant halt.

In practice, the phenomenon is referred to as wedging of self-locking gear and it is usually accompanied by considerable load causing in some cases damage. It must be mentioned that conditions of dynamic wedging differ from those of the static one.

3.1 Experimental Examination of Mechanical Systems

More obsolete methods of experimental examination of oscillation of toothed gears rested in monitoring of changes of torsional moments affecting the rotating shafts. Transfer of relatively weak signals from rotating shafts caused difficulties which were solved by means of telemetric transfer, i.e., by fixing a transmitter onto rotating shaft with consequent monitoring of the signal by a static antenna. To prevent disturbing effects in operation the signal modulated with frequency was used for transfer.

Modern methods of experimental examination stem in measuring of response of structures (drives) in their controlled actuation in the selected network of points on the surface. The assessed frequency transfers between the individual pairs of points serve for the determination of modal frequencies and regressive calculation on the basis of measured data is used for the determination of modal shapes of oscillations. The procedures are referred to as experimental identification or modal testing.

To illustrate oscillation shapes the graphical programs are used with animation of the individual oscillation shapes being activated gradually on the screen.

3.2 Mathematical Modeling of Drive with Cylindrical Worm Gear

Under the influence of load, the individual segments of worm gear get deformed and take new position. With respect to the fact that the contact of faces of teeth in the case of cylindrical worm gear systems occurs along twisted curve (contact line), it is not possible to predict load along the lines. Therefore, a simplification shall be used, which has already been applied, and load of segments of worm gear shall be concentrated in the pitch point C (see Fig. 5.3).

By implementation of coordinate system O(x,y,z) defined can be the magnitude of forces in the direction of individual coordinate axes applying load both onto worm (full lines) and worm wheel (interrupted lines). Their magnitude depends on the following:

- Torsional moment M_{k1}
- Angular speed ω_1
- Friction angle ρ

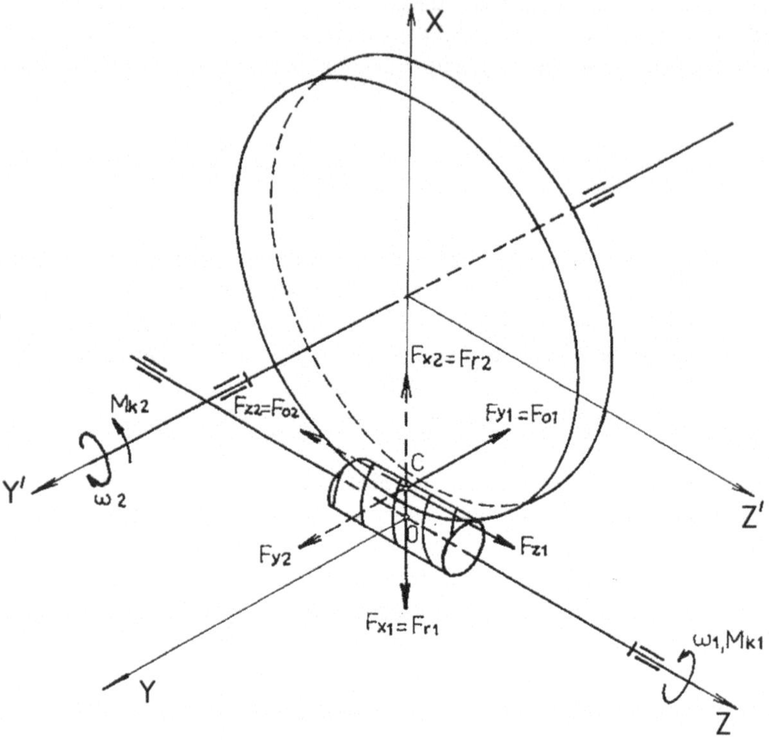

Fig. 5.3 Load of cylindrical worm gear

$$\text{with} \quad \rho = arctg \frac{f}{\cos \alpha_n}, \tag{5.2}$$

with

f—friction coefficient
α_n—angle of mesh in the standard plane [deg]

With respect to the aforementioned simplification (i.e., concentration of load into three mutually perpendicular forces) there is no point in a closer examination of load of teeth in the standard plane and therefore the attention shall be shifted to force acting in the direction tangential to pitch circle of worm wheel F_{o2} ($F_{z2} = F_{z1}$).

The term additional dynamic force of cylindrical worm gear system shall refer to increment of the force. Scalar equation is as follows:

$$F_{dz} = F_{zmax} - F_z, \tag{5.3}$$

with

F_{dz}—additional dynamic force [N]
F_{zmax}—maximal force acting in tangent to pitch cylinder of worm wheel [N]

$$F_z = F_{z2} \tag{5.4}$$

where F_{z2} is the force corresponding to transferred M_k [N].

According to Fig. 5.3 the following relations are applicable:

Circumferential force of worm = axial force of worm wheel:

$$F_{o1} = F_{y1} = F_{y2} = \frac{2M_{k1}}{d_1}, \tag{5.5}$$

with

M_{k1}—torsional moment of the worm shaft [Nm]
d_1—diameter of pitch cylinder of worm [m]

Circumferential force of worm wheel = axial force of worm:

$$F_{o2} = F_{z2} = F_{z1} = \frac{F_{o1}}{tg(\gamma + \rho)}. \tag{5.6}$$

Radial force of worm wheel = radial force of worm:

$$F_{r2} = F_{z2} = F_{z1} = F_{o2}\frac{tg\alpha_n \cos\rho}{\cos(\gamma + \rho)} = F_{o1}\frac{tg\alpha_n \cos\rho}{\sin(\gamma + \rho)}. \tag{5.7}$$

3.3 Mathematical and Physical Dynamic Model of Drive with Cylindrical Worm Gear

Assumptions of Solution

In the formation of dynamic model of drive, it is inevitable to take into consideration particularities of cylindrical worm gear. Figure 5.4 shows torsionally oscillating model of a drive with cylindrical worm gear.

Meaning of the individual symbols:

k_1—resulting stiffness of supports of worm in the direction of coordinate axes x, y [Nm^{-1}]:

$$\frac{1}{k_1} = \frac{1}{k_{p1}} + \frac{1}{k_{o1}}, \tag{5.8}$$

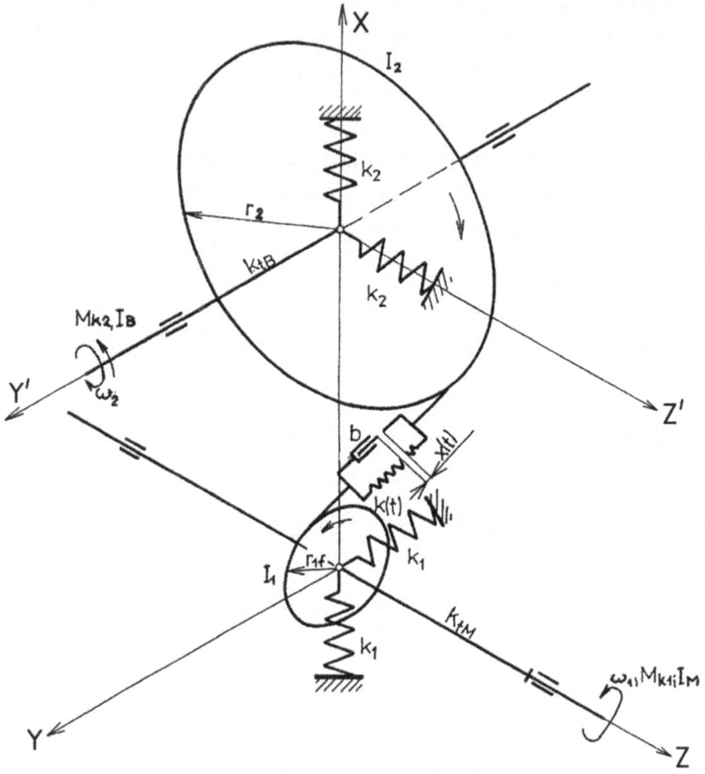

Fig. 5.4 Torsionally oscillating dynamic model of a drive with cylindrical worm gear

with

k_{p1}—gross stiffness of worm supports in the center of shaft distance [Nm^{-1}]
k_{o1}—flexural stiffness of worm shaft in flexure [Nm^{-1}]
k_2—resulting stiffness of supports of worm wheel [Nm^{-1}]
$x(t)$—deviation (fault) of toothing [m]
b—damping constant of toothing [Nsm^{-1}]
$k(t)$—variable stiffness of toothing [Nm^{-1}]
r_{1f}—assumed (fictitious) diameter [m]:

$$r_{1f} = r_1 tg\gamma, \tag{5.9}$$

with

r_1—diameter of pitch cylinder of worm [m]
γ—lead angle of pitch cylinder [deg]
r_2—diameter of pitch circle of worm wheel [m]
k_{tM}—torsional stiffness of worm shaft from part of driving motor [$Nmrad^{-1}$]

k_{tB}—torsional stiffness of shaft of worm wheel from part of loading (braking) moment [Nmrad^{-1}]
I_M—axial moment of inertia of rotor body of motor [kgm^2]
I_B—axial moment of driven machine inertia [kgm^2]
I_1,I_2—axial moments of inertia of bodies of worm and of worm wheel [kgm^2]

After reduction is performed in case of worm shaft it must be assumed that reduced average stiffness of toothing corresponds by order to torsional stiffness of worm shaft from part of driving motor k_{tM}. The fact does not allow acceptation of expectation of torsionally isolated gearbox. Yet the reduced torsion stiffness of worm wheel shaft k_{tB} per worm shaft is frequently by one order lower which allows accepting certain simplification.

In the first detailed description torsionally oscillating dynamic model of a drive with cylindrical worm gear as shown in Fig. 5.4 shall be examined and the following assumptions shall be made:

1. For all standardly produced worm gearboxes it can be assumed that both worm shaft and worm wheel shaft are symmetrical as to shape in relation to the center of shaft distance.
2. Worm and worm wheel are mounted in the center of shaft distance.
3. The first detailed description does not take into consideration flexural oscillations of worm and of worm wheel which can be reasoned by the fact that weight characteristics of worm are practically negligible and deflections of shaft of the worm wheel are by order lower with regard to its higher stiffness in flexure.
4. It is supposed that deviations of toothing with tooth frequency are lower contrary to static deformations and that bouncing of teeth faces shall be avoided and that flank clearance shall not occur.
5. Final strength capacity of supports in the direction of coordinate axes x and y is identical as in Fig. 5.4.

Coefficient of Mesh Duration

One of the most significant stiffness parameters in the dynamic calculation of toothed wheels is the coefficient of mesh duration on the basis of which mean stiffness of toothing is determined. Coefficient of mesh duration offers an idea of alternation of a single-, two-, and multi-pair mesh.

In case of spur-toothed wheels with straight teeth the coefficient of mesh duration is defined as follows:

$$\varepsilon = \frac{L}{p_b}, \tag{5.10}$$

with

L—length of meshing straight line limited by tip cylinders of pinion and globoid of wheel [m]
p_b—pitch of the basic cylinder [m]

$$p_b = p\cos\alpha, \tag{5.11}$$

with

p—pitch of the pitch cylinder [m]
α—angle of mesh [deg]

In case of spur-toothed wheels with skew teeth the coefficient of mesh duration *ε* is given by the total of mesh duration in the frontal plane and of mesh duration from step ε_k:

$$\varepsilon = \varepsilon_{\check{c}} + \varepsilon_k = \frac{L}{p_{\check{c}b}} + \frac{l_o}{p_{\check{c}r}}, \tag{5.12}$$

with

L—length of meshing line in the frontal plane [m]
$p_{\check{c}b}$—pitch in the frontal plane of the basic cylinder [m]
$p_{\check{c}r}$—pitch in the frontal plane of the pitch cylinder [m]
l_o—mesh length from step [m]:

$$l_o = b\ tg\beta, \tag{5.13}$$

with

b—width of the narrower of the wheels [m]
β—angle of inclination of teeth [deg]

Determination of mesh duration in the case of cylindrical worm gears is rather demanding especially due to the fact that the mesh does not occur along the straight line (plane surface) which is typical for spur toothing. Meshing area is a distorted surface and meshing lines are curves. Therefore, simplification is preferred, i.e., replacement of worm by gear rack with skew teeth. With regard to certain approximation in the determination of coefficient of mesh duration of cylindrical worm gears a more detailed analysis of the parameter was approached.

Numerical Method of Determination of Meshing Area and of Coefficient of Mesh Duration

Progress in computation technique allows precise determination of meshing field of worm gear systems and eliminates demanding factor connected with analytical and graphical examination.

The method is applied to determine meshing and contact line in parallel to axial cutting based on the statement that at the moment of mesh the central as well as the contact points appear to be lying on the joint standard of teeth faces. The statement is applicable also in case of planes parallel with the axial plane of the worm.

Generated and debugged programs for calculation in case of 11 parallel cuttings (number of cuttings is randomly selected) allow calculation in 29 points of meshing lines.

Program for determination of the length of meshing line numerically detects which point of meshing line in the respective parallel cutting represents the point falling into the category of points creating meshing field.

The analysis is performed for particularly given parameters of specific cylindrical worm gear. The term of category of meshing field points refers to points being connected into space limited to the following:

- Tip cylinder of worm
- Plane surfaces with mutual distance equaling to the width of a rim of worm wheel
- Largest external tip cylinder of the rim of worm wheel
- Globoid of the rim of worm wheel

Should such a point of meshing line meet the conditions, the program shall displace the straight line between the nearest point D_{n-1} still not lying in the meshing field and point D_n (Fig. 5.5). The solution shall allow alleviation of the step (according to required precision) along with consequent determination of the beginning of meshing field in case of respective meshing line with random precision.

Similar procedure is also applied in the determination of end point of meshing field "U." The length of meshing line occurring in the meshing field is determined out of which, consequently, coefficient of mesh duration in cutting j is specified as follows:

$$\varepsilon_j = \frac{L_j}{\pi m_x \cos \alpha_j}, \tag{5.14}$$

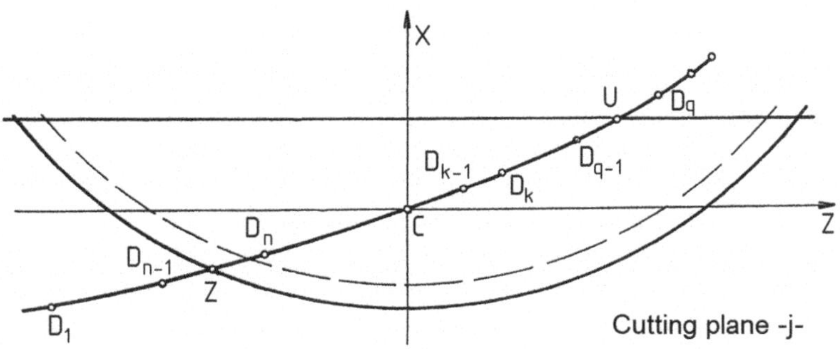

Fig. 5.5 Meshing area and coefficient of mesh duration

with

m_x—axial modulus of worm [m]
L_j—length of meshing line in cutting j [m]:

$$L_j = \overline{ZD_n} + \overline{N_n D_{q-1}} + \overline{D_{q-1} U}, \tag{5.15}$$

where

$$\overline{N_n D_{q-1}} = \sum_{k=n+1}^{q-1} (X_k - X_{k-1})^2 + (Z_k - Z_{k-1})^2, \tag{5.16}$$

where α_j is the average angle of inclination of meshing line in cutting j which can be determined according to the following relation:

$$\alpha_j = \frac{\alpha_{ZN} + \sum_{n-1}^{q-1} \alpha_k + \alpha_{q-1U}}{q - n + 1}, \tag{5.17}$$

with

$$\alpha_k = arctg \frac{X_k - X_{k-1}}{Z_k - Z_{k-1}}, \tag{5.18}$$

$$\alpha_{ZN} = arctg \frac{X_n - X_z}{Z_n - Z_z}, \tag{5.19}$$

$$\alpha_{q-1U} = arctg \frac{X_U - X_{q-1}}{Z_U - Z_{q-1}}. \tag{5.20}$$

Maximal value of coefficient of meshing duration in the case of individual parallel cuttings is considered to be the standard one.

Stiffness of Teeth

The chapter revolves around determination of stiffness (yielding property) of teeth of cylindrical worm gear system in the direction tangential to the pitch circle of worm wheel [13], i.e., parallel with the axis of the worm.

Stiffness of cylindrical worm gear systems was the subject of research carried out by Klimenko and Romanenko [18] who in case of calculation apply the method of final elements and perform experiments with the models made of organic glass. With regard to the fact that the available literature lacks information on stiffness characteristics of teeth of cylindrical worm gear system, the calculation of yielding property of worm gear system shall contain relations for spur gear system with skew teeth in accordance with [13]. At the same time yielding property occurring due to shaft flexure or due to yielding property of worm supports shall be taken into consideration.

(a) Yielding property from tooth deflection as of a built-in beam:

$$\delta_1 = 1{,}45 \cdot 10^{-11} E', \tag{5.21}$$

with δ_l being the yielding property from tooth deflection as of a built-in beam correlating with the unit of width of worm wheel for a single pair of jointly meshing teeth [m^2N^{-1}],with

$$E' = \frac{\frac{E_{Fe}}{E_1} + \frac{E_{Fe}}{E_2}}{\cos\gamma}, \tag{5.22}$$

with

E_{Fe}—modulus of elasticity in tension of steel [Pa]
E_l,E_2—modulus of elasticity in tension of used material of worm and of worm wheel [Pa]
γ—lead angle [deg]

(b) Yielding property from deformation of surfaces of teeth faces:

$$\delta_2 = 1{,}45 \cdot 10^{-10} E' \left(0{,}75 - 0{,}115 \log \frac{1{,}45 \cdot 10^{-4} F_{o2} z_2}{B p_n \cos^2 \gamma} \right), \tag{5.23}$$

with

δ_2—yielding property from deformation of surfaces of teeth faces per unit of width of worm wheel [m^2N^{-1}]
F_{o2}—circumferential force of worm wheel [N]
z_2—number of teeth of worm wheel
B—width of worm wheel [m]
p_n—pitch in the standard plane [m]

(c) Yielding property from shift of teeth of worm wheel with attention paid to the wheel body:

$$\delta_3 = \frac{1{,}45 \cdot 10^{-12}}{\cos\gamma} \left(\frac{E_{Fe}}{E_2} + \frac{0{,}8 p_n}{a_2} \cdot \frac{E_{Fe}}{E_2} \right), \tag{5.24}$$

with

δ_3—yielding property from shift of teeth of worm wheel with regard to the body of wheel per unit of width of worm wheel [m^{2}N^{-1}]
a_2—radial thickness of rim below the root circle of worm wheel [m]

Final yielding property of 1 pair of teeth from deflection, contact deformation, and shift is given by

$$\delta_v = \sum_{i=1}^{3} \delta_i. \tag{5.25}$$

Consequently, unit stiffness of teeth refers to

$$k_I = \frac{1}{\delta_v}, \tag{5.26}$$

with k_I being the yielding property of 1 pair of jointly meshing teeth per unit of width of worm wheel [Nm^{-2}].

Further on, given are the orientation values of unit stiffness of toothing from flexure, contact deformation, and shift during meshing of 1 pair in case of gearboxes UCG80 (smallest size), UCG160 (medium), and UCG315 (largest), the producer of which is the company of ZTS Sabinov, Slovakia:

$$k_{\mathrm{IUCG80}} = 1{,}124 \cdot 10^{10}\,\mathrm{Nm}^{-2},$$

$$k_{IUCG160} = 1{,}153 \cdot 10^{10}\,Nm^{-2},$$

$$k_{IUCG315} = 1{,}017 \cdot 10^{10}\,Nm^{-2}.$$

The following can be written in case of stiffness of two pairs of teeth during meshing:

$$k_{II} \approx \frac{k_I}{0{,}6}, \tag{5.27}$$

And for meshing duration of $2 < \varepsilon < 3$ in case of skew teeth of spur gear systems the ratio of stiffness values is as follows:

$$\frac{k_{II}}{k_{III}} \approx 0{,}75, \tag{5.28}$$

with k_{II}, k_{III} being the stiffness of 2 or 3 pairs of teeth.

Consequently, for mean stiffness of meshing with meshing duration of $2 < \varepsilon < 3$ the following can be written:

$$k_o \approx \frac{k_I}{0,45}(\varepsilon - 2) + \frac{k_I}{0,6}(3 - \varepsilon), \tag{5.29}$$

with k_o being the mean stiffness of cylindrical worm gear system from flexure, contact deformation, and shift [Nm^{-2}].

With regard to the unavailability of decent basic documents for the determination of mean stiffness of toothing of cylindrical worm gear taken into consideration are values measured in case of spur gear systems with skew teeth, i.e., the worm is replaced in the first detailed description by skew gear rack.

3.4 *Mathematical and Physical Dynamic Model of Cylindrical Worm*

In the formation of dynamic model of worm gearbox, it is not possible to take into consideration torsional isolated gearbox as for instance in [15, 16].

Contrary to spur-toothed gear system, it causes higher gear ratio, lower mean stiffness of toothing, and mainly considerable flexural yielding property of worm shaft. The yielding property which in the highest degree contributes to overall yielding property of flexibly deforming elements of gear is transformed to torsional stiffness of gear to stiffness in the direction tangential to the pitch circle of worm wheel. In the case of majority of cylindrical worm gears (UCG160) after reduction and formation of reduced torsional system it shall be detected that reduced mean stiffness of toothing is lower or correspond by order with torsional stiffness of worm shaft from part of the driving machine k_{tM}. However, the reduced torsional stiffness of shaft of worm wheel k_{tB} from part of the driven machine is mostly at least by a single order lower contrary to the reduced stiffness of toothing:

$$k = \frac{1}{\frac{1}{k_o B} + \frac{1}{k_{oh1}} + \frac{1}{k_{oh2}}}, \tag{5.30}$$

with

k—mean stiffness of toothing [Nm^{-1}]

k_o—mean stiffness of toothing of cylindrical worm gear from flexure, from contact deformations, and from shift of teeth per unit of width of wheel [Nm^{-1}]

B—width of worm wheel [m]

k_{oh1}—flexural stiffness of worm shaft transformed to stiffness in tangent to pitch circle of worm wheel [Nm^{-1}]

k_{oh2}—flexural stiffness of shaft of worm wheel in tangent to pitch circle of worm wheel and parallel with the worm axis [Nm^{-1}]

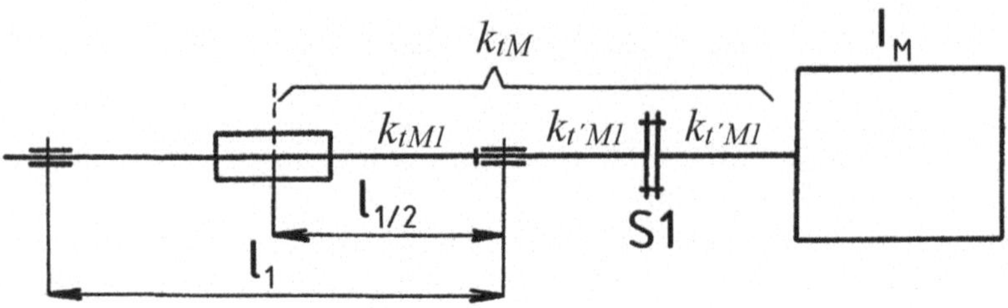

Fig. 5.6 Torsional stiffness of worm shaft

Torsional stiffness according to Fig. 5.6 is considered to be torsional stiffness of worm shaft from part of driving motor k_{tM}:

$$k_{tM} = \frac{1}{\frac{1}{k_{tM1}} + \frac{1}{k_{t'M1}}}, \tag{5.31}$$

in case of which k_{tM1} is the torsional stiffness of worm shaft from the center of bearing distance up to bearing from part of driving motor [Nmrad^{-1}].

$k_{t'M1}$ is the torsional stiffness of worm shaft from bearing up to clutch S_1 or from clutch to rotor of driving motor I_M [Nmrad^{-1}].

It can be assumed that clutch S_1 is absolutely stiff and torsional stiffness of shaft of driving motor up to its rotors is (in case of certain simplification) equal to the stiffness of worm shaft from bearing to clutch S_1.

The same is applicable for torsional stiffness of shaft of worm wheel k_{tB} as in Fig. 5.7:

$$k_{tB} = \frac{1}{\frac{1}{k_{tB2}} + \frac{2}{k_{t'B2}}}, \tag{5.32}$$

in case of which k_{tB2} is the torsional stiffness of shaft of worm wheel from the center of bearing distance up to bearing from part of the driven machine [Nmrad^{-1}].

$k_{t'B2}$ is the torsional stiffness of shaft of worm wheel from bearing to clutch S_2 or from clutch S_2 to I_B [Nmrad^{-1}].

On the basis of the analysis of stiffness parameters of cylindrical worm gear the following mutual relations were detected:

1. Due to higher gear ratio, which is typical for worm gear, after reduction per shaft of worm rather low reduced axial moment of inertia of the driven machine is achieved.
2. Reduced mean stiffness of toothing is lower or corresponds by order to torsional stiffness k_{tM}.

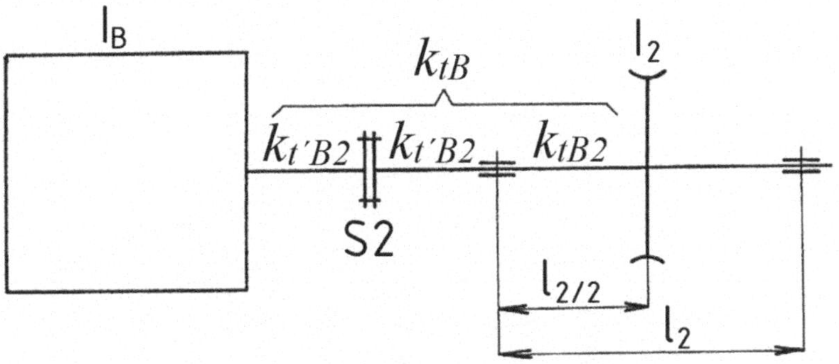

Fig. 5.7 Torsional stiffness of shaft of worm wheel

3. Reduced mean stiffness of toothing is higher contrary to reduced torsional stiffness k_{tBred} or in case of average and larger sizes the difference is expressed in orders.
4. With regard to reduced axial moment of inertia of worm wheel I_{2red}, I_M is higher by orders [17].

To get oriented the values of calculated stiffness for average size of gearbox UCG160 (axial modulus $m_x = 6.3$ mm, $\alpha = 20°$, number of runs $z_I = 4$, transference number $u = 10$) shall be given:

$k_{tM} = 1.124\ 105$ Nmrad^{-1}, $k_{tred} = 3.177\ 104$ Nmrad^{-1}, $k_{tBred} = 4.379\ 103$ Nmrad^{-1}, in case of which k_{tred} is the reduced torsional stiffness of toothing [Nmrad^{-1}]:

$$k_{tred} = \frac{kr_{1f}^2}{u^2}, \qquad k_{tBred} = \frac{k_{tB}}{u^2}, \tag{5.33}$$

3.5 *Influence of Geometrical Parameters Upon Inherent Frequencies*

By means of computing program which uses the method of impedance matrix mentioned the inherent frequencies of dynamic model of cylindrical worm drive were calculated as well as of dynamic model with worm drive. Detected was the influence of axial moment of inertia of worm wheel and of flexural stiffness of worm shaft upon change of inherent frequency of cylindrical worm drive.

The results are plotted in spatial graph (Fig. 5.8) for the monitored gearbox. The axial moment of inertia as well as flexural stiffness was being changed within the range from 0.6 to 1.4 times with regard to currently produced state. The influence is clear from the given figure.

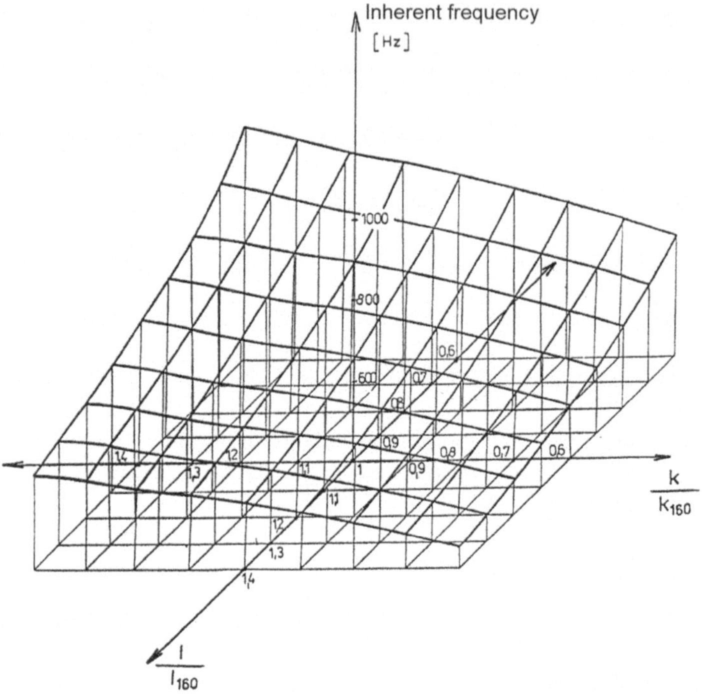

Fig. 5.8 Gearbox UCG160: the axial moment of inertia as well as flexural stiffness depending on the inherent frequency

3.6 *Dynamic Load of Teeth of Cylindrical Worm Gear*

Figure 5.8 shows the development of values $\overline{\eta_1}$ for the UCG160 gearbox. Value K' ranges from 0.0858 to 0.0835 for the individual gearbox types. If simplification of $I_l \approx 0$ is accepted (weight moment of worm is not taken into consideration) and after inclusion of variable stiffness of toothing into series with weight from part of motor, K' is achieved within the range from 0.0178 to 0.014, i.e., influence of variable stiffness becomes negligible.

Influence of External Actuation

Programs for dynamic analysis allow calculation of deviation, speed, or acceleration in any random node of dynamic model. In a similar way it allows determination of force or moment in any random element of the model [18] (Fig. 5.9).

$$\text{Mesh duration} = 2,707, b_p = 0,3$$

Development of additional dynamic force was determined by the implementation of external sinusoid unit force and by the detection of value of additional dynamic force in toothing at random frequency of actuating force. Development is shown in Fig. 5.10 for dynamic model. The values of relative damping are indicated in the respective graph.

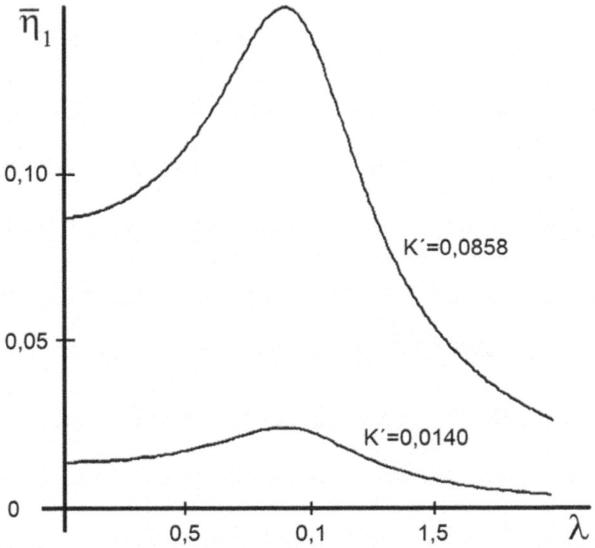

Fig. 5.9 Development of values: Gearbox UCG160

$$\text{UCG160}: 1-b_p=0,1, 2-b_p=0,2, 3-b_p=0,3, 4-b_p=0,4$$

4 Conclusion

Contrary to generally accepted opinion on the absence of dynamic phenomena in worm gears this chapter provides detailed analysis of the meshing process of teeth of worm wheel with worm threads and whether dynamic load is possible in worm gears as well.

It has been proved that in the case of reverse gears and drives operating in the mode of frequent start-ups and halts, the impacts can occur in the result of which lubricating medium is forced out of contact zone of wheel teeth and worm threads and that leads to metal contact and further on to abrasion and wear.

It has been shown that to alleviate and to prevent the impact the elastic adjacent element should be located among the teeth of wheel teeth and worm threads which would serve as a damper of impact system and at the same time it shall assure minimal coefficient of friction during relative slip of work areas of worm threads and teeth of worm wheel. The character of collision of worm threads and wheel teeth depends on the type and properties of elastic adjacent element among wheel teeth and worm threads.

Reduction of vibration activity of worm gears which increases their service life can be achieved by use of modern original structures of worm wheels in which elastic elements eliminating dynamic load get integrated. Polymer films on work areas of worm threads and teeth of worm wheels or cluster lubricant, which is standard lubricant with added Teflon particles, can serve as elastic adjacent element.

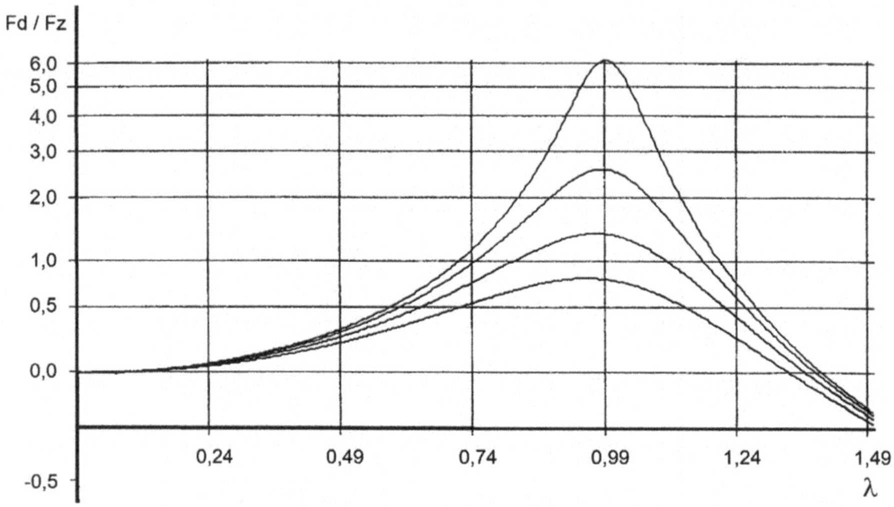

Fig. 5.10 Development of additional dynamic force acting in toothing in case of external actuation and diverse damping

Acknowledgement Presented research has been supported by the projects KEGA 051TUKE-4/2018 and VEGA 1/0823/21. This work was also supported by the Slovak Research and Development Agency under contract No. APVV-18-0316.

References

1. Smeringaiova A, Wittner M (2018) Parametric modeling as efficient way of designing. MM Sci J (December):2701–2705
2. Milacek S(1990) Modal analysis method and its application, 1st edn, ČSVTS Prague, Machine construction cycle, 136, Prague
3. Panda A, Olejárová Š, Valíček J, Harničárová M (2018) Monitoring of the condition of turning machine bearing housing through vibrations. Intl J Adv Manufact Technol 97(1-4):401–411
4. Straka L, Dittrich G(2020) Influence of MTP on surface roughness and geometric accuracy of machined surface at WEDM 2020. Intl J Eng Manage Sci:91–97
5. Saga M, Vasko M, Pechac P (2014) Chosen numerical algorithms for interval finite element analysis. Procedia Eng 96:400–409
6. Husar J, Knapcikova L (2019) Exploitation of augumented reality in the industry 4.0 concept for the student educational process. In: INTED 2019: Valencia (Spain): IATED, pp 4797–4805
7. Gaspar S, Pasko J (2018) Mathematical formulation of the kinematic equations for the control of the robot system with application for the machining conical surfaces. MM Sci J (March):2158–2161
8. Klimo V (1977) Kinematic-dynamic characteristics of engagement and their influence on the bearing capacity of tooth surface of cylindrical worm gears. Habilitation thesis, VŠT Košice
9. Balazikova M (2016) Assessment of ergonomic risks of the selected work activity by niosh method. Sci Bus Soc 1(5):42–45
10. Vejc VL (1974) Dinamičeskije charakteristiky mašinnogo agregata s samotormozjaščimsja mechanizmom. Zubčatyje i červjačnyje peredači, Mašinostrojenije

11. Vejc VL (1965) Dinamika samotormozjaščichsa červjačnych peredač pri silach trenija, zavisjaščich ot skorosti, Teoria mašin i mechanizmov, Vyp. 105–106, M. izd-vo Nauka
12. Vejc VL (1968) Dinamika mašinnogo agregata a samotormozjaščejsa peredačej pri učete zazorov v zacepleniji. Zubatyje i červiačnyje peredači, Mašinostrojenije
13. Tuplin AW (1972) Gearing stress. SNTL, Prague
14. Zeman V, Němeček J (1991) Mathematical modeling and modal synthesis of systems with spur gears, 42
15. Olsansky O(1978) Gear oscillation with variable gear stiffness. Mech Eng 28(6)
16. Zeman V (1971) Dynamic loading of spur gears, Mech Eng 21
17. Coranic T, Gaspar Š Pasko J (2019) Applications of 3D printing in mould structures. RAM-Verlag, Lüdenscheid, 85 p
18. Klimenko AA, Romanenko FA (1974) Izgibnaja žestkosť zubjev koles cilindričeskoj červjačnoj peredači. Sb. Detali mašin, Technika

Chapter 6
Research into the Impact of Modern Marketing Strategies on the Competitiveness of the Company and Increasing the Economic Potential of the Company

Annamária Behúnová, Marcel Behún, and Ervin Lumnitzer

1 Introduction

In its beginnings, marketing appeared only in more advanced economies. We can deduce from the literature that marketing first appeared in the twentieth century. However, its foundations go deep into history—from the time when the markets themselves began to exist. Marketing prevailed in schools in 1902 when marketing began to be taught mainly at American universities. Marketing came to Europe a little later, around 1950. In Europe, people perceived it in all sorts of ways, but later they began to accept it and thus became a part of our daily lives. Marketing has not only changed the way we do business in America and Europe but also spread to other world economies [1, 2].

Marketing development can be divided into four basic stages [1, 2]:

- Latent stage—a period that lasted from the beginning of the market mechanism until the beginning of the twentieth century, when marketing began to develop. Absence of theoretical foundations, because this period was characterized mainly by the natural marketing behavior of traders.
- The stage of establishment of a saturated market—a stage that was closely related to the great economic crisis (at the turn of the 1930s and 1940s). A period crowded with competing companies that did not have a chance to succeed in

A. Behúnová (✉) · M. Behún
Institute of Earth Resources, Technical University of Košice, Košice, Slovakia
e-mail: annamaria.behunova@tuke.sk; marcel.behun@tuke.sk

E. Lumnitzer
Faculty of Mechanical Engineering, Department of Process & Environmental Engineering, Technical University of Košice, Košice, Slovakia
e-mail: ervin.lumnitzer@tuke.sk

L. Knapcikova et al. (eds.), *Sustainable Management of Manufacturing Systems in Industry 4.0*, EAI/Springer Innovations in Communication and Computing,
https://doi.org/10.1007/978-3-030-90462-3_6

such a crowded market. The period of introducing the planning and use of marketing into business practice.

- Marketing management stage—the 1950s and 1960s of the last century. During this period, the dynamic development of the most advanced economies in the world took place. There was also the development of new technologies, increased competition, and increased production. New concepts such as marketing control, marketing planning, and marketing implementation are being introduced.
- Stage of marketing strategic management—the last and most developed stage, which currently operates in many markets around the world. For this reason, today's companies have such goals: originality (in terms of marketing and marketing strategies), effort to expand their operations at the international level (larger market), education and training of employees, constant market analysis, strategic marketing, and efforts on longer term cooperation.

1.1 Strategic Marketing

Chandler defined strategy as defining the most basic goals in the long run and then adapting the behavior to these defined basic goals. This mainly means the accumulation of resources that are used to adapt the company to its environment [3].

Schendel and Hatten discussed strategy and strategic marketing in the company in more detail. They argue that the strategy in the company can be understood as the basic goals of the company, procedures for achieving these goals, as well as ways of accumulating resources in the company. They also support the view that these accumulated resources are used to adapt the company to its environment [3].

In today's hectic times, modern societies are forgetting the importance of strategic thinking and planning. This is mainly due to the lack of time in newly established companies. Smaller companies in particular make a huge mistake. They think that the process of planning and strategic thinking only applies to large companies, but the opposite is true. According to P. Kotler, these smaller companies are planning their end. He argues that if a company plans and thinks ahead, it can better rework its goals and principles. This leads to better efficiency, which results in the increased performance itself. Thanks to planning, the company can anticipate developments or react to them immediately, because the company is ready for such changes [1].

Strategic marketing takes into account especially the long-term period. Strategic thinking and planning are also closely connected with this strategic marketing. It is strategic thinking that helps ordinary employees understand the importance of planning, and also helps to implement the very decisions made by the company's strategic management (see Fig. 6.1) [5]. In the case of strategic planning, companies in most cases prepare the following plans [4]:

The annual plan is a short-term plan of the company, the task of which is to describe the current state, goals of the company, program and budget for the given year, as well as strategies.

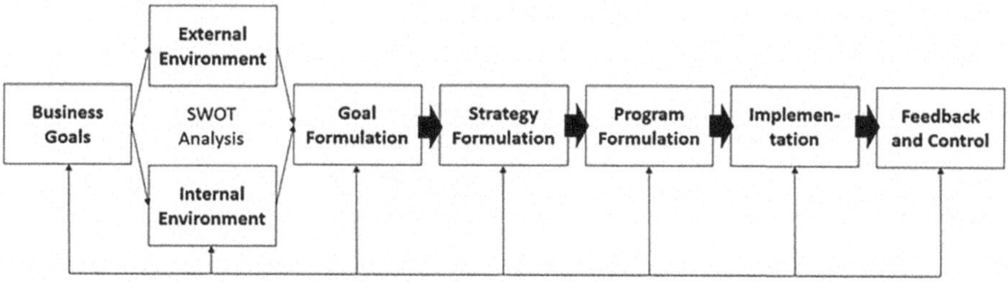

Fig. 6.1 Strategic planning process [4]

Table 6.1 The difference between a long-term plan and a strategic plan [2]

Long-term plan	Strategic plan
It is planned periodically.	It is planned continuously.
The current trends are expected to continue.	Changes are expected that will affect society.
The situation "in what the company does business" is addressed.	The situation "in what the company should do business" is addressed.
Planned from behind C, B, A.	Options are planned.

The long-term plan defines the main factors and forces that will affect the company over the next few years. Such a long-term plan contains the company's long-term goals, primary strategies, and resources needed to achieve these long-term goals. The long-term plan is adjusted every year due to timeliness.

The strategic plan deals mainly with the ability to take advantage of the opportunities that companies present. They are changing in a constantly changing environment. The strategic plan also seeks to maintain a balance between the company's goals and capabilities (see Table 6.1).

2 Marketing Strategy of the Company

Strategic marketing and marketing strategy are the two terms that we come across quite often. However, both of these concepts are different, and therefore we can characterize them as follows. The term strategic marketing includes all theoretical knowledge that is closely related to the strategic planning itself, which takes place in a market environment. Strategic marketing is a set of all known knowledge that we can use to our advantage. Under the term marketing strategy, we must imagine something completely different. A marketing strategy is a specific strategic plan in a particular company that works according to pre-agreed conditions and agreements. The marketing strategy is closely related to the company's goals [6].

In recent years, there has been enormous technological progress and development, which has logically had to be reflected in the economy, marketing, and marketing strategies themselves. These changes occur mainly in the external marketing environment, which also results in a modification of the marketing mix, which was

Table 6.2 Transformation of marketing mix 4P to 4C [6]

4P	4C
Product	Customer value
Price	Cost to customer
Place	Convenience
Promotion	Communication

most often mentioned and implemented by companies as the so-called 4P. However, based on the changes in the company, criticism has begun to emerge that the 4P marketing mix only concerns the perspective of the company and not the consumer, thus modifying it to 4C (see Table 6.2).

The transformed marketing mix looks like this [7]:

- The product becomes a customer value.
- The price changes at the customer's cost (cost to the customer).
- The place changes to customer convenience (convenience).
- Marketing communication changes to communication with the customer (communication).

Thanks to this, we find that from the customer's point of view, value, low price, great comfort, and communication are important, i.e., no promotion. 4P thus becomes 4C.

At present, marketing strategies focus mainly on the customer, i.e., the consumer, who is the most important part of the so-called new economy. It is the needs of customers that are given priority and companies place great emphasis on them in planning, decision-making, and subsequent control.

Most companies today think that if bad times come for a company, there is a need to reduce costs and the area that will limit marketing the most. However, if a company decides to limit marketing, it leaves room for all its competitors. The competition acquires new market shares at the expense of such companies and thus strengthens its market-competitive position. During financial, economic, and other crises, customers lose their reasons for shopping. Unless the company itself gives them such reasons, its sales may fall rapidly, even at a time when the market situation is calming down and improving. Current trends have also been responsible for changes in the thinking of managers in creating new marketing strategies.

2.1 *Aftermarketing*

Aftermarketing can be described as a set of steps that a company takes after completing the sale of a product or service to maintain the loyalty of its customers in resale. For products that need service support, such as cars, after-sales service is very important and also forms a major part of aftermarketing. In traditional marketing, the main emphasis is on the sale of the product or service itself, while in aftermarketing, marketing strategies start only after the sale of such services or products takes place. The basic pillar of this marketing is to build a relationship between the seller and the buyer, which the company strives to retain the customer in the future. Here it is true that the sale is not the end, but on the contrary, it is the beginning of the relationship between the seller and the buyer [8].

Aftermarketing consists of seven activities [9]:

- Creation of a database of customers (current, former, inactive but also future)
- Identification and characteristics of interactions with their customers
- Examining the feedback of their customers (satisfaction, dissatisfaction)
- Customer satisfaction survey (consumer interest survey)
- Informing customers
- Organizing and creating various events and programs for customers
- Review, reach, and recover your old lost customers

2.2 E-marketing

We can rank e-marketing among the most modern trends in marketing, which have huge potential. In this area, we can include activities from the sale or purchase of products and services online to information and communication with customers. In terms of market supply, we can divide companies into two basic groups. Pure-click companies (PCCs), which were created on the Internet and never existed before, and brick-and-click companies (BCCs), which in turn existed as classic stone companies, but due to new trends and developments began to present on the Internet as well [10].

For PCC, the beginnings were very difficult because people did not trust online shopping. The reason was also the fact that most of the companies that sold over the Internet did not have any brick-and-mortar stores and were essentially unknown to their customers. Initially, commercial sites sought to sell books, music, and toys. At the end of the 90s of the last century, there was a huge expansion of these companies, especially in the USA. Some companies have shown much higher profits, such as the United Airlines or Pepsi. Around this time, BCC companies began to adapt to the new trend, gradually starting to create websites. However, these companies feared that there would be a conflict of interest, as their online store itself was theoretically to compete with their stone shop [10].

This method of advertising brings several benefits for all parties involved. When setting up the campaign itself, the advertiser can already set how much money he or she is willing to spend for the given campaign. Also, such an advertiser pays only for the advertisement used (after clicking on the link). For a potential customer, this has several advantages, especially that the advertisement is displayed based on his or her interests and is therefore not so annoying. From the point of view of the owners of the websites on which these ads appear, the biggest advantage is that the site generates a certain profit for its operation, which is often many times greater than the cost of such a website (this is especially true for various blogs, magazines, news websites, etc.). Last but not least, the company that places this ad on all websites earns the most from such advertising and at the same time takes care of the very operation of this entire system. The downside can be a lot of competition for specific keywords. The advertiser must set a higher cost-per-click for those keywords, which may not pay off in the end. That is why such ad campaigns are thoroughly analyzed, set up, reanalyzed, and edited [11].

2.3 Guerilla Marketing

Guerrilla marketing can be classified into unconventional forms of marketing. Levinson defined guerrilla marketing as early as 1982 as an unconventional way to reach future customers at the lowest possible cost. The goal is to attract potential customers in such a way that they do not know that it is some promotion of something. The success of guerrilla marketing is based on the tactics of gaining new people (consumers, co-workers), weakening competition and survival in the market [10].

For the success of guerrilla marketing, it is necessary to choose the right tactics, which must include at least three predefined areas for the proposed communication campaign. The first area is associated with an attack in an unexpected place, which means that there is a choice of a place where the entire campaign will take place. After determining the location, the focus is on precisely defined goals. These are selected customer segments, distribution points, etc. Another logical step that follows from guerrilla marketing is withdrawal, as guerrilla marketing is mostly one-off events [12].

Effective guerilla marketing consists of the following six attributes [10]:

- The use of a nontraditional channel that consumers try to avoid.
- This method is usually cheap.
- Guerrilla marketing uses psychology to its advantage.
- Binds secondary publicity.
- In a more aggressive variant directly attacks the competition.
- We can put practically any activity into guerrilla marketing.

After guerrilla marketing, most companies do not have sufficient financial resources to directly fight their competition. The basis of this marketing is an original idea with an even more creative design. In the beginning, this strategy was used by smaller companies due to a lack of funds. Nowadays, however, the card has been reversed and guerrilla marketing is being used by larger and larger companies [10].

2.4 *Mass Customization*

One of the relatively new marketing strategies is mass customization. This strategy originated in the USA and aims to meet the specific needs of customers. The term was first used in 1987 by Staven Davis in his Future Perfect, in which he stated that it was a way of producing one type of product, based on customer specifications, regardless of the economics of production [13].

However, mass customization did not come to the fore until 1993, thanks to Pine, who defines it in his publication "Mass Customization: The New Frontier in Business Competition" as the ability to design and manufacture customized products with the same efficiency and speed of production process as mass production [14]. The main goal is for every single consumer to find exactly what he or she needs in the market of products and services, but without increasing the price. Some methods of customization offer the consumer the opportunity to directly influence some specifics of the final products; otherwise, the range of products offered is expanded and the level of satisfaction of consumer needs is increased through a greater choice of different final goods of one kind [[14], 15].

Customization methods [14, 15]:

- Creation of individualizable products and services
- Individualization of services for standardized products and services
- Taking into account individual requirements when delivering goods
- Modularization of components to meet individual requirements
- Achieving a fast response in the value chain

Some authors consider complex customization, custom localization strategy, scalability [16], or product customization [17] to be mass customization.

2.5 *Product Placement*

Product placement works on the paid placement of products (brands) directly into the content of a specific television program or medium. The product can be presented visually or verbally as part of the film scene. The popularity of this marketing strategy is due to the worldwide development of digital and information technologies. Thus, the consumer is not bothered by annoying advertising while watching his or her favorite content, but lurking on it directly while watching the content. Product placement affects mainly emotions, perception, and memory and subsequently can influence the consumer's decision-making during the next purchase. The most common product placement products are cars, branded clothing, food, beverages, and electronics [10].

2.6 *Relationship Marketing*

Relationship marketing is based on strengthening relationships with your consumers. With this marketing strategy, managers try to create a link that occurs between the consumer and the product. This is a consequence of the fact that they have begun to realize that satisfied consumers are behind the long-term success of the business [18].

Philip Kotler defined relationship marketing as the process of creating, maintaining, and expanding strong, valuable relationships with customers along with other stakeholders [19]. Every year, the average company loses about ten per cent of its customers, while only a five per cent reduction in the transition of customers to competitors can increase profits by up to 25-85%. It is more likely that if the company can keep the customer as long as possible, profitability will automatically start to rise [20].

In relationship marketing, we distinguish five levels of relationships that companies can build with their customers who have purchased a particular product or service from them [19]:

- Basic—The company sells its product and no longer cares about anything.
- Reactive—The company sells its product and at the same time asks its customer to contact the company in case of any problems or ambiguities.
- Responsible—After selling the product, the company contacts the customer and finds out whether the product meets the customer and meets his or her expectations. At the same time, the seller receives feedback from the customer, thanks to which the company can improve its product in specific areas. This information helps the company to constantly improve and build its market position.
- Proactive—The seller occasionally contacts his or her customer after the sale with questions about improvements or new products.
- Partner—The company, the seller, and the customer are in constant contact and cooperation to find out together how the company can provide better value to the customer.

2.7 *Word-of-Mouth Marketing*

This type of marketing can be included in the field of marketing, where the main task is to provoke the oral spread of "advertising" among the customers themselves. Word-of-mouth marketing can be supported through various promotional activities created by the company, or by hiring an external company for this work. In the last decade, the growth of this type of marketing has become very widespread, mainly due to the development of technology and media. The task of advertising is to acquaint the customer as clearly as possible with the benefits associated with the purchase of a particular product or service. The customer must be so interested that there will be an oral spread of good experiences with other potential customers [10, 21].

Buzz marketing: This is a specific form of marketing that causes the oral dissemination of information about a product or service. Buzz marketing aims to create a powerful experience that leads people to talk to each other naturally about the company, brand, and product. Buzz marketing is sometimes defined as getting positive reviews and recommendations, directly from customers. Its role is to engage consumers and the media to such an extent that writing and talking about a product, brand, and company will be very interesting, valuable, and exciting. Buzz marketing's main task is to cause a stir, to give consumers reasons to talk about the product with each other. This form of marketing is quite cheap but very difficult to prepare a marketing program. Six main features lead to the successful application of buzz marketing in practice—unusualness, content of the secret (or its disclosure), remarkability, humor, crossing the line, and breaking taboos (sex, lies, harsh humor). Buzz marketing also works with viral marketing [22].

Viral marketing began to emerge with the advent of the Internet. It consists of creating interesting, creative, and funny products (e.g., pictures, videos, applications), which are then forwarded to each other by e-mails, SMS messages, and social networks. Viral marketing is not very expensive and we know it in two forms: passive, where the company relies on the positive reaction of the consumer by offering him or her a quality product, and active, which works on the principle of a viral message that directly influences consumer behavior to increase product sales [10]. For a viral campaign to be successful, we must meet a few conditions: sharing the message must not be complicated; the message must be interesting, funny, or valuable; the object of interest must be free; and forwarding must not be associated with a reward—people should forward of your own free will. We can also use image, text, audio, and video for such a message. Even with more demanding campaigns, message dissemination must be as simple as possible. There is a high probability that the dissemination of the message will not only affect the target group but also penetrate the message into various other segments. Thanks to the Internet, viral marketing has been given unlimited possibilities, communication has been speeded up and simplified, and it is more playful and interesting. There is also direct feedback between the consumer and society. The effective use of viral marketing also occurs in connection with a competition or an interactive game [10].

2.8 Experiential Marketing

Experiential marketing does not promote a product just by acquainting the buyer with its features and benefits. It also connects him or her with unique and interesting experiences. Simply put, selling a product is not the most important thing. The main thing is to show the consumer that this product will enrich his or her life [23]. The representatives of this marketing strategy are Pine and Gilmore, who claim that we are currently on the threshold of a new era of the economy. Companies should thus strive to organize memorable events for their consumers [15].

The goal of an experiential marketing strategy is to create a kind of connection between the customer and the company. However, not all companies need the same type of connection. A skilled manager will first find out which type of connection best suits a particular company, which he or she will then use. There are eight basic types of connections [24], [25].

Emotional connection: Using experiential marketing to evoke emotions is also one of the possibilities in experiential marketing. Simply put, if a company brings its customers to tears, it will achieve what it needs. These experiences have the power to unite people. From social "events" they can connect different audiences, even individual cities. The emotional connection allows companies to cross the imaginary gap and be able to reach the customer's heart [24].

Educational link: In this link, information intended for education is considered to be the main factor. Training takes place within the company (training of internal employees), but also in external conditions. Then we talk about educating external customers and long-term clients about new products, use of services, and changes in society. Research has shown that if a customer understands and is familiar with a particular product, they are four times more likely to reach for that product [24].

A combination of surprise and pleasure: Seemingly random to the eye, but completely designed step by step. The basis is the use of the moment of surprise from which follows another connection. Kerry Smith judged from his study that a surprising experience is etched deep in his memory and has a long-term impact on the customer, much longer than traditional marketing [24].

Trial connection: The company will demonstrate part of its product, or have it tested by its customers. This also applies to any other service. We can also find a test connection outside on the street in the form of small samples, in retail to sophisticated testing of new technologies at trade and demonstration exhibitions [24].

Encouraging connection: Motivates the end customer to connect with a particular company. Sometimes a simple discount is enough. Other times it is a gift or bonus. Incentives can be small but also large and may not always have a monetary value. Some companies use to access bonus content or events as an incentive [24].

Motion connection: We are talking about using a joint effort to create a connection. According to many managers, this is the most ecological and "clean" type of connection. Often much more than just a link. Permanent links are created that last much longer than any other type of link [24].

3 Methodology

The creation and performance of a marketing strategy are a creative process in which the marketing strategy itself is formed. The number of steps can be larger or smaller, as the further division is governed by the rules of a particular company [6].

Determination of starting points: Sets clear starting conditions, which include in particular the answers to two basic questions. Who will be responsible and who will create the company's marketing strategy? What are the initial goals for creating and executing a company's marketing strategy [6]?

The second step is the analysis of the current marketing situation, which says that in the preparatory part, information was processed by market research, or was selected from the marketing information system. Deeper penetration into the issue is required through established system analysis. The tasks focus mainly on the characterization of the target market, competition, intended customer, and characteristics of the offered goods. Within the abovementioned tasks, it is recommended to proceed very slowly and carefully, especially with the help of qualified people [6].

The philosophical and conceptual direction is the third step, in which the direction of the company must be determined, which will be followed by a marketing strategy. However, we must not forget that it is the clarification and concretization of the philosophical and conceptual direction that is of great importance in the preparation of a marketing strategy [6].

Specifying the strategic position is the fourth step in which the responsible staff evaluates the basic goal together with the information obtained. Before starting to form the marketing strategy itself, it is important to conclude and solve two tasks—specifying strategic goals and specifying the problems, tasks, and pathways leading to these goals. The past position or positions of the company are taken into account, as well as the position that is a kind of vision of the future in the emerging marketing strategy [6].

The formation of a marketing strategy, from the point of view of a company operating in a market environment, is the most important step in the formation of a marketing company itself. From this moment on, the overall direction of the company develops. We are talking here in particular about the six steps, which are graphically illustrated in Fig. 6.2 [6].

The approval of the marketing strategy can also take place in the so-called simplified form. The draft marketing strategy is submitted to the top management. He or she can approve it and proceed with the implementation. However, the whole process can be different if the creation of the strategy did not take place under top management. In this case, however, there are often misunderstandings and

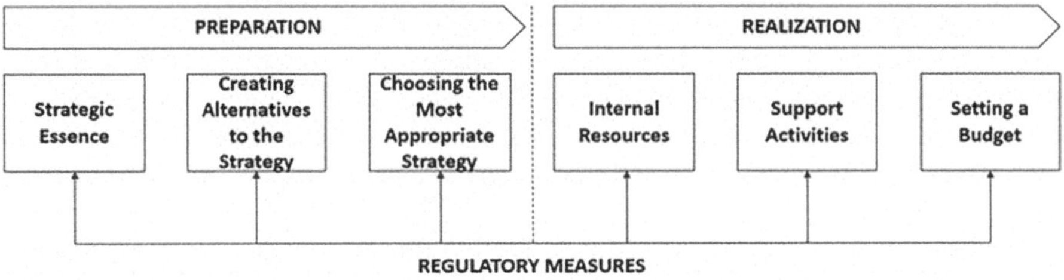

Fig. 6.2 Algorithm for shaping marketing strategy [6]

contradictions [6]. As approval affects the direction and existence of the company for a longer period, it should also be given increased attention in the formal area:

- Approval of the marketing strategy takes place at several levels (middle management, top management, etc.).
- The approval protocol is also used for approval, where more serious problems are noted, which are specifically recorded and approved.

If the company has determined the set plan (strategic, tactical, or operational), the implementation process begins, in which the planned and anticipated proposals are changed and integrated into marketing management practice [6]. The company's management must distinguish between the preparation phase, the implementation phase, and the control phase. These three control functions are often overlapped and complicated. Implementing plans into practice is managerial and costly but also very time consuming for the company [6].

The final step in creating a marketing strategy is to ensure control activities. There must be a feedback cycle between planning and control. The control process meets several attributes, the main ones being accuracy, timeliness, objectivity, continuity, and feedback. The control process must be performed at all levels of management, including operational control, tactical control, and strategic control [6].

4 Results

In 2008, the economic crisis, which affected not only Europe, also deeply affected the society in which the research was carried out. The low purchasing power of the population caused considerable problems for the company, but thanks to the tenacity of the owner, these problems succeeded in "stabilizing" these problems during 2008-2012 and the company maintained its existence and avoided bankruptcy. To avoid stagnation, in 2018, a new marketing strategy began to be considered—E-marketing, specifically BCC, i.e., e-commerce.

The development but also the temporary decline of the company is shown in Fig. 6.3, from which it is evident that the company's turnover in the years 2002–2008 had an upward trend. This positive situation lasted until the advent of the economic crisis, which, as mentioned above, was a major threat to the business environment not only in Slovakia but also globally. In the years 2008–2012, the company's turnover dropped rapidly and it could be said that the company shook its foundations. Fortunately, after the time of the crisis, a turning point occurred and society slowly began to prosper again. Thanks to the expansion of the product range, the company's turnover has taken an upward trend. As is obvious, the year 2018 and the establishment of the e-shop helped the company to increase its turnover to its current maximum (see Table 6.3).

The last month monitored was March. During this month, the company decided to try two modern marketing strategies:

- Product placement in the form of a paid PR article
- Viral marketing in the form of a funny picture

Fig. 6.3 Flow chart of the company's turnover in the years 2002 to 2018

Table 6.3 Monthly sales and the average value of orders for the period under review

A month	Number of orders	Total revenues	Average order value
September	15	427.30 €	28.49 €
October	19	615.40 €	32.39 €
November	34	1,024.70 €	30.14 €
December	48	1,458.60 €	30.39 €
January	23	1,257.40 €	54.67 €
February	74	3,147.60 €	42.54 €
March	41	2,180.20 €	53.18 €
Total	**254**	**10,111.20 €**	**39.81 €**

The PR article was a paid alternative, while viral advertising worked for free and its distribution was taken care of by users of the social network Facebook. During March, there was a slight decrease in orders compared to those in February. However, this was because the impact of paid advertising was much smaller than during the previous month. However, total sales were higher than during the pre-Christmas period in December. During March, the company received 41 orders with a total value of €2180.20.

From Fig. 6.4, it might seem to us that paid promotion with Google Ads was ultimately less successful than free promotion. However, we need to keep in mind that paid promotion with Google Ads only ran for one month, while free promotion has been running since the online store was launched in September, October, January, as well as partly in November and December. The same applies to paid promotion via the social network Facebook, which was launched 28 days.

From Fig. 6.4 at the same time, we deduce that during the 7-month operation of the online store, the company managed to obtain 36% of orders through free promotion and up to 64% of orders thanks to paid promotion via Facebook, Google Ads, PR article, and viral advertising. In Table 6.4 we can see the number of individual orders for specific sources of acquisition.

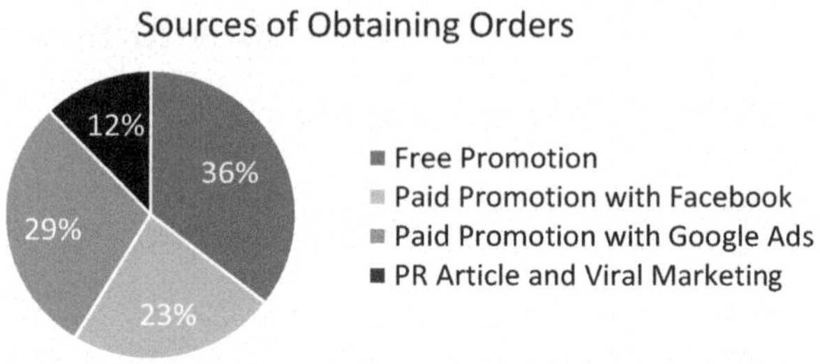

Fig. 6.4 Sources of obtaining orders during the period considered

Table 6.4 The number of orders and their percentage per source

Source	Number of orders	% share
Free promotion	90	36%
Paid promotion with Facebook	59	23%
Paid promotion with Google Ads	74	29%
PR article and viral marketing	31	12%
Total	**254**	**100%**

Table 6.5 The costs and revenues for individual sources of marketing

Source	Costs	Revenues
Creation and operation of an online store	240 €	3,640.03 €
Paid promotion with Facebook	100 €	2,325.58 €
Paid promotion with Google Ads	100 €	2,932,25 €
PR article and viral marketing	100 €	1,213.34 €
Total	**540 €**	**10,111.20 €**

However, if we look at the orders we received more specifically, we find that out of the total number of orders, 90 were obtained through free promotion. The company received another 74 orders through paid promotions using Google Ads. In the third place was paid promotion via the social network Facebook with the number of orders 59. In the last place is paid product placement (PR article) together with viral marketing. These two modern marketing strategies were launched before the end of the reporting period and together won 31 orders (see Table 6.4).

The total cost of creating, launching, and promoting the online store for 7 months was €540. However, the costs of creating and running an online store are minimal, as the company decided to create an online store on its own, without the help of an external specialized company. €240 represents the annual costs associated with renting the domain and hosting on which the online store itself is stored. These costs for the creation and operation of the online store are paid once a year in advance. The remaining €300 was used to promote the online store to gain new customers (see Table 6.5).

5 Conclusion

In today's modern and fast world, almost every company that offers some products for sale should be online. These companies, which we also call a brick-and-click company, must also start offering their products through online stores; otherwise, it may be that traditional companies are outnumbering the newly established Internet "pure-click companies." This was not the case with the company under investigation, which also suffered losses due to competition, precisely due to online shops. The company thus had to expand its operations from a small region to the entire Slovak Republic.

The surveyed company invested €540 in its online store and generated sales of €10,111.20, through orders, the total number of which for the period under review was 254. 254 orders do not mean only 254 customers across Slovakia. Ultimately, these customers can influence their acquaintances when shopping. From that moment on, the company operated not only regionally, but also nationwide, because the company obtained orders from various parts of the Slovak Republic. In a way, the company has become a more competitive company on the market thanks to e-commerce.

From these findings, we know that an online store cannot run without paid advertising at this time. This is especially true for newly opened online stores, which customers do not yet know about, and online stores must build a name and brand from scratch. However, paid advertising is also beneficial for functioning online stores, which have been on the market for several years. They expand their offer among new people and thus gain potential customers. If companies want to succeed in the oversaturated market or maintain their competitiveness, they must focus their attention on reaching potential customers thanks to various tools of modern marketing.

References

1. Kotler P (2007) Moderní marketing. Grada, Praha
2. Matúš J, Ďurková K, Čábyová Ľ (2012) Moderný marketing. Księży Młyn Dom Wydawniczy, Lódž
3. Lesáková D (1990) Strategický marketing. Profis, Bratislava
4. Kotler P, Keller KL (2013) Marketing management. Grada, Praha
5. Kafka T (2005) Dovednosti řízení podniku. Český institut interních auditorů, Praha
6. Jedlička M (1998) Marketingová stratégia podniku. Magna, Trnava
7. Rozbor marketingového mixu. http://www.robertnemec.com/marketingovy-mix-rozbor/. Last accessed 28 Dec 2020
8. Aftermarketing. http://www.mbaskool.com/business-concepts/marketing-and-strategy-terms/10800-after-marketing.html/. Last accessed 28 Dec 2020
9. Vavra T (1995) Aftermarketing: How to Keep Customers for Life through Relationship Marketing. Irwin Professional Publishers, Chicago
10. Hesková M, Štarchoň P Marketingová komunikace a moderní trendy v marketingu. Oeconomica. Praha. ISBN 9788024515205

11. Kontextová reklama. http://marketing.krea.sk/clanky/kontextova-reklama/. Last accessed 28 Dec 2020
12. Guerilla marketing. http://www.marketingovenoviny.cz/marketing_4039/. Last accessed 11 Jan 2021
13. Davis SM (1987) Future perfect. Addison-Wesley Publishing, Reading
14. Pine BJ (1993) Mass customization The new frontier in business competition. Harvard Business School Press, Boston
15. Pine BJ, Gilmore JH (1993) The experience economy. Harvard Business School Press, Boston
16. Logman M (1997) Marketing mix customization and customizability. Business Horizons 40(6), pp. 39+.
17. Åhlstrom P, Westbrook R (1999) Implications of mass customization for operations management: an exploratory survey. Intl J Operat Prod Manage 19(3):262
18. Keller K (2007) Strategické řízení značky. Grada, Praha
19. Kotler P (2003) Marketing management. Prentice-Hall, Upper Saddle River
20. Reicheld FF (1996) The loyalty effect. Harvard Business School Press, Boston
21. Word of Mounth marketing. http://www.investopedia.com/terms/w/word-of-mouth-marketing.asp. Last accessed 11 Jan 2021
22. Rosen E (2002) The anatomy of buzz: how to create word of mouth marketing. Crown Publishing, New York
23. Post P, Brand B (2000) The power of experience branding. ANA/The Advetiser
24. Smith K, Hanover D (2016) Experiential marketing: secrets, strategies, and success stories from the world's greatest brands. Wiley, Hoboken
25. Mesaros P et al (2020) The impact of information and communication technology on cost reducing in the execution phase of construction projects. TEM J 9(1):78–87

Chapter 7
Use of Marker AR by Implementation on Production Equipment for Education of Students of Technical Fields: A Study

Jozef Husár, Jakub Kaščak, and Stella Hrehová

1 Introduction

In real world in which the world of imagination, unreal and often influencing human understanding and thinking, increasingly interferes. Our life is increasingly affected by computer, imaginary reality, which enters our consciousness either completely in an imaginary, virtual form, but in a form mixed with the real world. Therefore, the aim of this chapter is to point out the possibilities of using augmented reality in two forms. One is the creation of markers that can be implemented in the technical documentation (sketch) and also directly in the production equipment.

1.1 Mixed Reality

It is possible technically to create the so-called mixed reality by combining the real and virtual worlds. Mixed reality (MR) can be understood as a blending of the real world with virtual reality. Elements of both the real and virtual scenes in various proportions are entered into mixed reality. Mixed reality can be technically applied in two dimensions [1] (Fig. 7.1).

J. Husár (✉) · S. Hrehová
Department of Industrial Engineering and Informatics, Faculty of Manufacturing Technologies, Technical University of Košice, Prešov, Slovak Republic
e-mail: jozef.husar@tuke.sk; stella.hrehova@tuke.sk

J. Kaščak
Department of Computer Aided Manufacturing Technologies, Faculty of Manufacturing Technologies, Technical University of Košice, Prešov, Slovak Republic
e-mail: jakub.kascak@tuke.sk

L. Knapcikova et al. (eds.), *Sustainable Management of Manufacturing Systems in Industry 4.0*, EAI/Springer Innovations in Communication and Computing,
https://doi.org/10.1007/978-3-030-90462-3_7

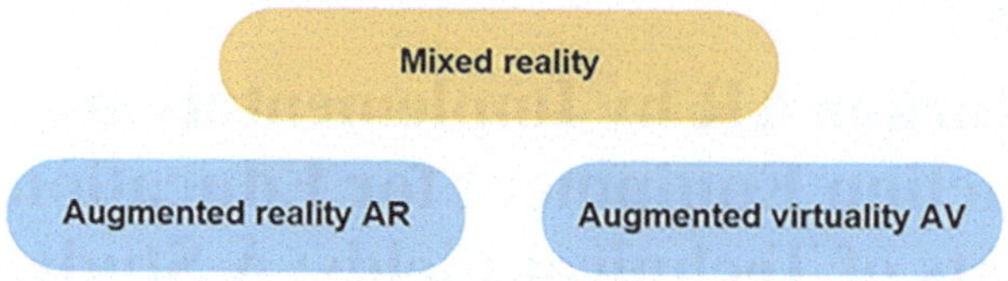

Fig. 7.1 Categorization of mixed reality

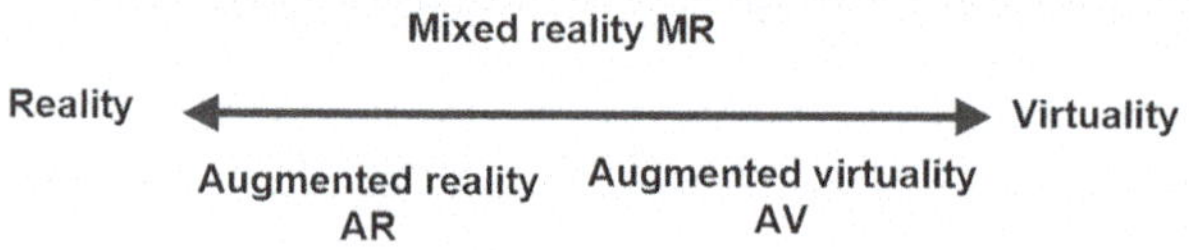

Fig. 7.2 Categorization of mixed reality—Milgram continuum [3]

Mixed reality is characterized [2] as

- A combination of real and virtual worlds
- Interactivity
- Processing in real time
- Registration in three dimensions

Augmented reality (AR) and augmented virtuality (AV) are quite similar. The difference lies in the fact that while in augmented reality, virtual features are applied to the real world, in augmented virtuality, it is the opposite. Here, real objects are inserted into virtual reality. The relationship among MR, AR, VR, and reality is characterized in Fig. 7.2.

1.2 Augmented Reality

In augmented reality, we can understand a certain transition between computer simulation and natural perception. The number of virtual elements can be various and ideally these elements should be indistinguishable from the real ones [4]. The graphic relationship between the real and the virtual environment describes the so-called Milgram continuum (Fig. 7.2). The augmented reality system is definable by a set of interconnected elements. Thanks to these connections, the system is characterized in relation to its surroundings. It gains specific properties, other than individual parts of the system without interconnection. The augmented reality system is classified into four main parts (Fig. 7.3).

By environment we mean the real environment to which the virtual object will be applied. The user is the target group for whom the application will be intended. By platform we mean the technical side of the application rather than the device, or the operating system of the end device, or other requirements within the application

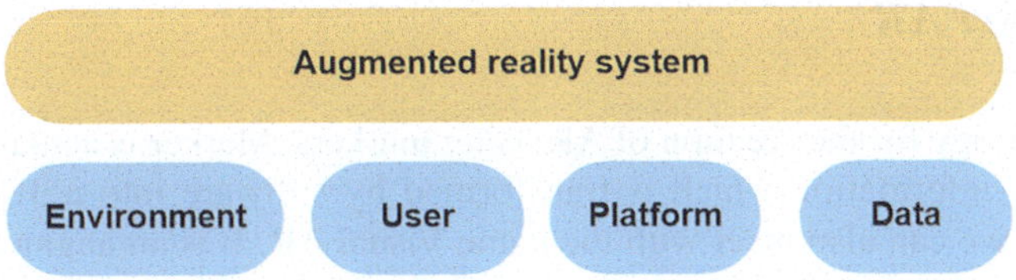

Fig. 7.3 Parts of augmented reality system [5]

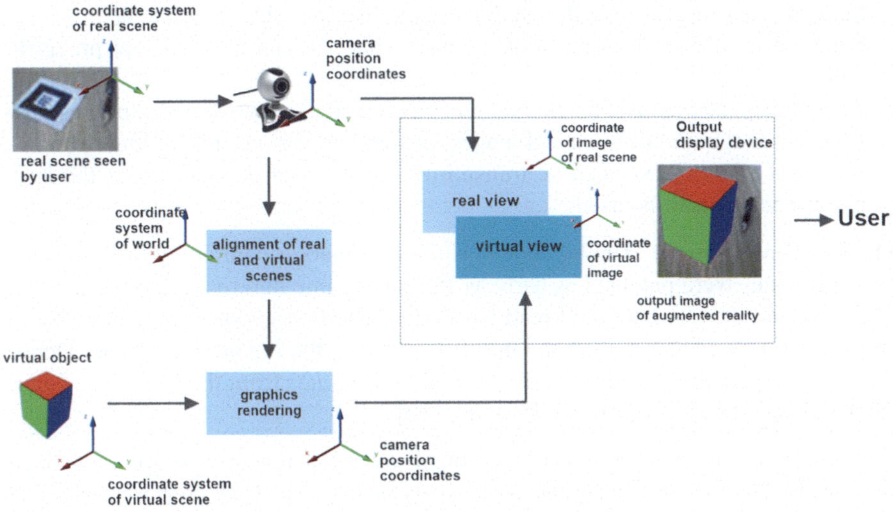

Fig. 7.4 The process of creating content in augmented reality [3]

development. The data characterize a precisely determined point of a virtual element in a real scene. It can be a description, a link, or other information.

The basis of augmented reality is to combine a virtual image with what the user sees. The starting point is the real scene that the user sees. The coordinate system of the real scene is captured by a camera, from which the real scene is transmitted to the output device based on the coordinate system, and the coordinate system for the virtual scene is sent in the 2D plane and processed by calculating the resulting virtual graphic scene. The graphics are rendered and the result as a virtual object is transferred to the output device. The generation of the image of virtual reality is realized by a graphical computer system, for example, based on OpenGL, where this system needs to collect information about the real scene, in terms of the location of the output virtual object [6]. Finally, in the output, the real image is merged with the virtual object. This process is illustrated in Fig. 7.4.

1.3 Marker AR

It is a technology for the creation of AR using markers. Marker is actually a sign of some digital information which is transformed by software into a 3D object. At Marker AR, we can also meet with the name Visual AR. It is an augmented reality that is oriented in space by recognizing a predefined point. This point can be of various natures, such as black and white markers, which, when recognized, are replaced in real time by a virtual model, text, web page, and other [7, 8].

Characteristic properties of markers:

- Accuracy: Devices respond quickly and accurately to the graphics of the marker, the user can rotate and tilt the model and the like in a 3D environment.
- Space: The marker does not require much space, it can be small but precisely defined.
- Availability: Nowadays, the marker can be created in a few seconds and it is possible to see it commonly on web pages, posters, on various advertising items.
- Variability The marker can be transmitted, so the user decides where the augmented reality will be used.

Marker Development The barcode was the first marker commonly used in practice. Bar codes were patented as early as 1952 and came into real use only in 1974. There was no device that could read bar codes. The first goods ever scanned were chewing gum at a supermarket in Ohio [9]. In 1994, the Japanese company Denso Wave developed a two-dimensional QR code. QR comes from the Quick Response label and was developed for fast decoding [10].

Another widely used group is printed black and white markers, which are defined very easily. Natural markers create a relatively new group. They are based on a real image or photo. As the last group of markers, we will mention real markers. Here, the marker can be any real object. There are several ways to create a 3D object. One way is to create a completely new 3D object using software applications [11] (Fig. 7.5).

Implementation of Marker AR During the implementation of AR and implementation in practice, the process goes through several phases. One of the phases is the creation of a 3D object, which will be displayed in a real scene after implementation. There are several ways to create a 3D object. One way is to create a new 3D object with an appropriate software application. Another way is to scan the object with a 3D scanner and the last option is to edit the 3D model. The next phase of creating the Marker AR is checking, correcting the 3D model, and then exporting it to a usable format. These formats are usually VRML, OBJ, 3DS, and others. The next phase concerns the selection or creation of a specific marker to which the above-mentioned 3D object will be assigned in the next phase. The penultimate but important part of the whole process of creating augmented reality is to assign with the help of peripherals and application equipment the 3D object to a precisely determined marker. The last phase is the resulting output of augmented reality to the

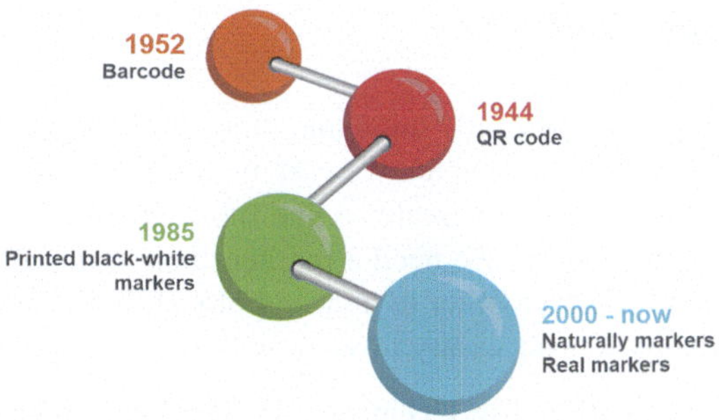

Fig. 7.5 Development of Marker AR

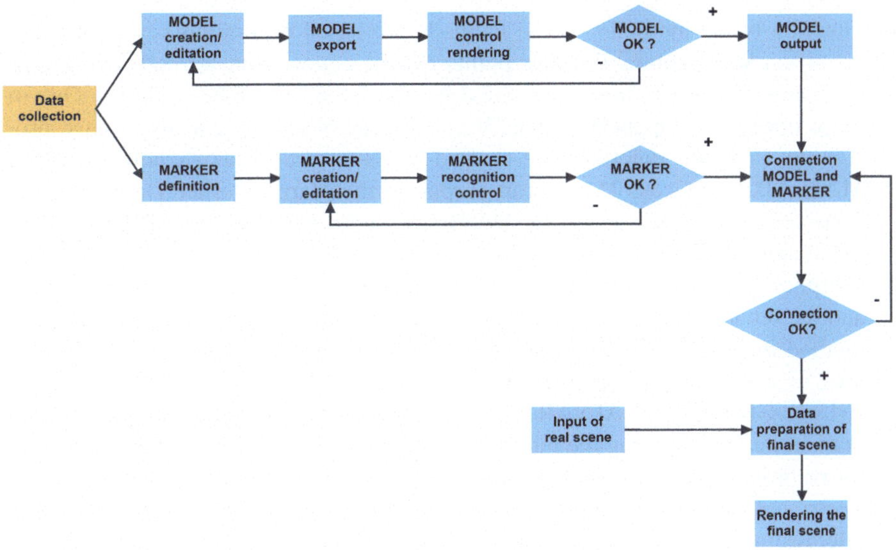

Fig. 7.6 The process of creating augmented reality using a marker

output device. The process of creating an augmented reality Marker AR is shown in Fig. 7.6 [3].

The biggest problem of Marker AR is the markers' creation. The markers should be well visible, precisely defined, and in contrast to being clearly recognizable, whether by a camera, mobile device, or other sensors. Today, scanning devices have different resolutions and at low resolutions the marker may not be loaded correctly. It means that the 3D object will not be assigned to the marker using application and eventually the augmented reality system will be inoperable [12].

1.4 Software

There are currently more and more applications for the creation of augmented reality. In this subchapter, we will mention at least some. First of all, we need to realize what we need to create so that the augmented reality system could be designed. In the beginning, we need a marker, then a 3D object, and finally an application that can assign a marker to a virtual object. It follows that in the creation of augmented reality it is necessary to work with two to three different applications.

Application for Creating 2D Marker The marker can already be created, or within the originality, we can create it ourselves. Today, there are a lot of generators freely available on the Internet for creating markers. Division according to marker creation technology is described below [13].

- Function-based methods: These methods are based on finding key points in a given marker graphic. The key points can be edges, corners, and others. The detection of these points is provided by a software-predefined library. One of these libraries is OpenCV. OpenCV is a library based on real-time image processing. Program functions are created in C++, it also has Python, Java, and MATLAB interfaces. This library supports the Harris Corner Detection, Shi Tomasi Corner Detector, FAST Algorithm for Corner Detection, BRIEF, ORB, SIFT, and SURF methods.
- Morphological methods: There we recognize two methods, the match template and the binary code. Programming software dealing with the matching template method includes ARToolKit and ATOMIC AuthoringTool, based on finding an area of a similar pattern directly in the marker. The programs Metaio SDK, ArUco, GRAFT, and ALVAR are characterized for encoding numbers in the marker, i.e., by applying the binary code method. We can use a 2D graphics vector or bitmap program, such as Adobe Photoshop, Adobe Illustrator, Inkscape, Vectr, and others, to create markers. The image resolution of 150x150 pixels after testing is sufficient at a distance of 15 cm from a regular laptop camera.

Applications for Creating 3D Objects These applications are important for creating a 3D model that will be applied to a real scene. They are, for example, [14]:

- **Sketchup** has an intuitive interface and easy access in terms of modeling objects. 3D models can be created in minutes. The application is also used by many experienced users, especially architects, who often create complex buildings and interiors in it. The library of add-ons and 3D model warehouses, created by our own users, is also helpful.
- **Fusion 360** combines not only computer-aided design (CAD) but also machining (CAM), structural analysis (CAE), and team collaboration. The application benefits from more than 30 years of Autodesk experience in the field of design software as well as a rich stock of tutorials (in English). It brought a fresh wind to design not only by its versatility but also, for example, possibility using calculations in the cloud.

- **Blender** is probably the most functionally equipped 3D modeler, which is developed under the open-source license—and is thus available to all users on Windows, Mac, or Linux platforms for free. Artistically oriented modelers, who do not want to be limited by a technical approach, but on the contrary appreciate free modeling in space, including the preparation of textures and advanced visualizations with sophisticated scene lighting, will find something to their liking here. Although the software is not primarily focused on creating surface-closed models for additive manufacturing, many advanced users cannot afford it.

Augmented Reality Implementation Applications Applications for the creation of augmented reality are mainly based on the ARToolKit platform and respective ARToolKit software library. ARToolKit uses algorithms that solve the main development problem, namely determining the position of the user's field of view. Software libraries for video surveillance without time delay calculate the position of the camera together with its relative orientation with respect to the scanned, predetermined object. Characteristics of ARToolKit [3]are

- Simple environment
- Libraries on platforms (Windows, MAC, SGI, Linux)
- Overlapping of the 3D model to place a real marker
- Support for peripheral inputs and various formats
- GUI interface
- Fast marker detection
- Fast 3D rendering
- Java as MATLAB support
- Modular API in C language
- Complex of examples and utilities
- OpenSource with GPL license for noncommercial use

Unity 3D

This software product, in a simplified sense, combines several elements into one. Inputs can be models, textures, block diagrams of the game, and last but not least, ideas. Dynamics and characteristic features are added to each element applied to the program via JavaScript, C #, or less used Boo. This system produces 2D and 3D applications and games from simple to complex. Unity is a multiplatform application, which means that its outputs can be used in Windows, MAC OS, Linux, and also mobile operating systems. The application development environment is called the UnityScript editor, whose comfort is ensured by color-coded coding, a spectrum of error messages when creating incorrect code strings, which facilitates work in the Unity environment itself [15].

Vuforia

The platform is an environment for creating applications with your own reality. It is a tool for easy creation of augmented reality and in conjunction with another program can be a powerful 3D tool for future applications. Vuforia was tested in conjunction with Unity 3D using Software Development Kit, but there were complications in terms of compatibility and the 64-bit Unity 3D engine platform.

Vuforia is compatible with C ++ programming languages and Java which allows the application of products for Android, iOS, and others. Vuforia is widely known as a leader in industrial augmented reality through its best-in-class computer vision technology, robust tracking capabilities, and wide platform support. Today, it is the most widely used software for handheld devices and digital glasses with 600K+ registered developers and hundreds of leading global enterprise customers across 30+ verticals—no other technology is used in multiple applications to deliver augmented reality performance [16].

BuildAR

It is a software application that allows us to create simple augmented reality scenes on the desktop. Augmented reality (AR) is a way to interact with the real world and virtual objects at the same time. Three-dimensional computer graphics are overlaid with the real world in a way that makes them part of the real environment. Creating an AR experience presents technical challenges and requires a variety of technologies, including video capture, image processing, 3D math, and computer graphics. BuildAR provides a graphical user interface that simplifies the AR scene creation process, allowing us to create the first augmented reality on the desktop. Everything that we need are a computer, a webcam, and some printed designs [17].

ARTag

It is an augmented reality support system. It can be used to facilitate the appearance of virtual objects, games, and animations within the real world. Like the previous ARToolKit system, it allows video viewing, which calculates the position and orientation of the camera with respect to physical marks in real time. Once the camera's position is known, a virtual camera can be positioned at the same point, revealing the virtual object at the location of the ARTag. It therefore solves two key problems in augmented reality, namely the viewpoint tracking and virtual objects interaction [18].

2 Methodology

The model for the application of augmented reality is the lathe PD 400/CNC, which is used for precise turning of steel and nonferrous metals. The lathe can be used for tiling and longitudinal turning, turning of balls, radii, and all free contouring of steel and nonferrous metals. Machining of the workpiece is performed automatically by software [24].

Technical parameters:

- Power supply: 220–240V / 50/60Hz
- Drive: Powerful condenser motor with 550W
- Recirculating ball spindle for the X-axis with 2.0 mm inclination and side diameter of 8 mm. Stepper motor with 1.8A and 50N/cm
- Travel distance: approx. 70mm

- Recirculating ball spindle for the Z-axis with 4 mm inclination
- Side diameter of 12mm. Stepper motor with 1.8A and 50N/cm
- Six spindle speeds: 80–160–330–660–1400–2800/min
- Stepper motor control: CNC image of the control unit (included)
- Software on CD-ROM, installation under Windows 98,2000, XP, Vista, and 7
- Drive connection: via RS 232 interface (or use of USB adapter), connection cables to PC included in the scope of delivery

Dimensions:

- Machine: length 900 x width 400 x depth 300 mm
- Control unit: length 450 x depth 270 x width 60 mm
- Total weight: Machine: approx. 45 kg / Control unit: approx. 4kg

2.1 Work Methodology

The basis for creating an augmented reality lathe will be its 2D scheme, which in the next work will serve as a scanned real area (Fig. 7.7).

We will use lathe pictures as well as a lathe description to create augmented reality. As part of the software applications, we will work with Adobe Photoshop CS3, Blender 2.77, and BuildAR PRO 2.2.1. In the first step, we will work with Adobe Photoshop CS3, where we will process the images and at the same time we create our own IRQ codes. We will create six IRQ codes. Of which four codes will be used to display images (parts of the lathe), one code will display the description, and one IRQ code will display the 3D-created lathe module in augmented reality. IRQ codes will be in a resolution of 150 x 150 pixels. After creating the IRQ codes, we will place these codes on the 2D drawing of the lathe. In the second step, we will create a 3D lathe module using the Blender 2.77 application and export it in OBJ format. Here, we will start with the picture of the lathe (Fig. 7.8).

In the last step, we will create augmented reality in the BuildAR PRO 2.2.1 application. We will import all IRQ codes and assign either a picture, a description, or an object to each code.

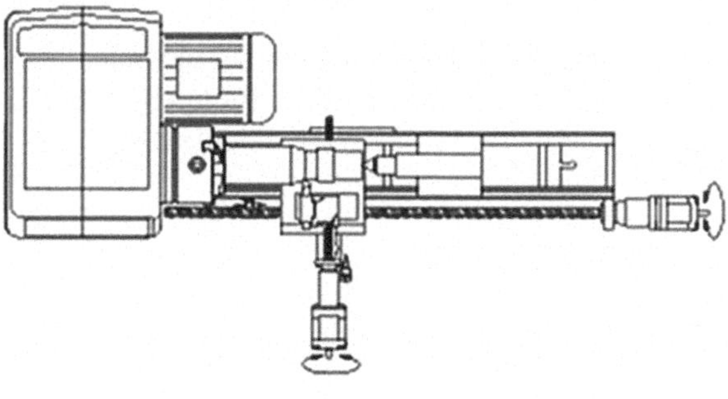

Fig. 7.7 Lathe PD 400/ CNC scheme

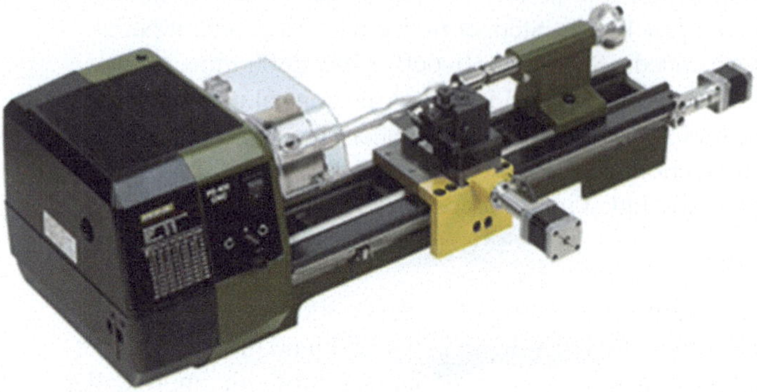

Fig. 7.8 Lathe PD 400/CNC picture

3D VIEW FOTO 1 FOTO 2

FOTO 3 FOTO 4 DESCRIPTION

Fig. 7.9 IRQ markers codes

Created IRQ Codes and Their Distribution The IRQ codes were created by the bitmap application mentioned above and they are shown in Figs. 7.9 and 7.10.

We created the 3D module in the Blender application and exported it in the OBJ format, which is supported by the BuildAR PRO 2.2.1 application, where we will assign a specific IRQ code to the module (Fig. 7.11).

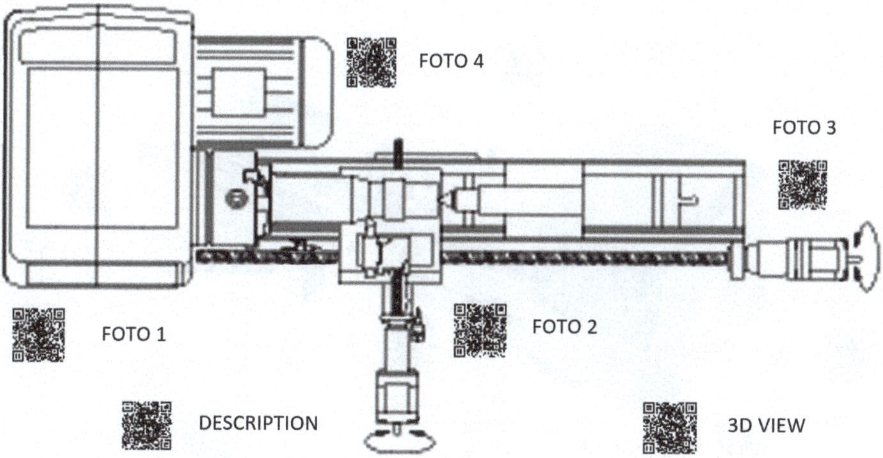

Fig. 7.10 Layout of IRQ codes in a 2D drawing

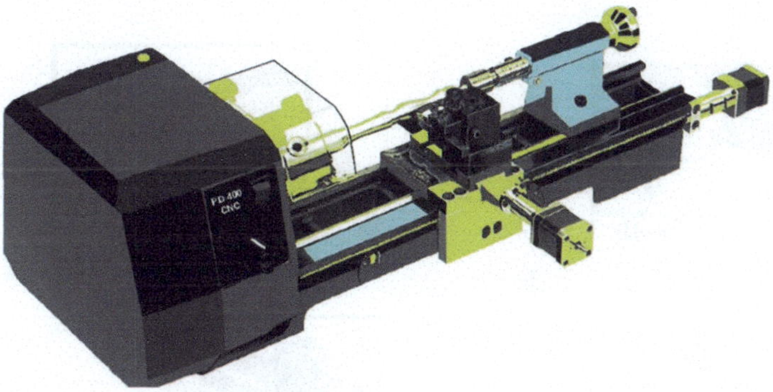

Fig. 7.11 3D lathe model

Assignment of the required outputs to the created marker and the output of augmented reality

In this final part, we have imported all six markers (IRQ codes) in the BuildAR PRO 2.2.1 application and we have assigned the required output information to each code. We can test the augmented reality directly in the BuildAR application, where we capture a 2D drawing with a laptop camera or a camera connected to a PC and the required outputs are displayed on the screen after scanning the IRQ codes with the camera. The expected outputs of the display are shown below (Figs. 7.12 and 7.13).

Such an augmented reality design can be created using the above-mentioned but also other appropriate applications. It can be tested directly in the application of assigning the required outputs to the imported marker with the help of a camera and an output device, a display. The output can be text, image, 3D object, and also video.

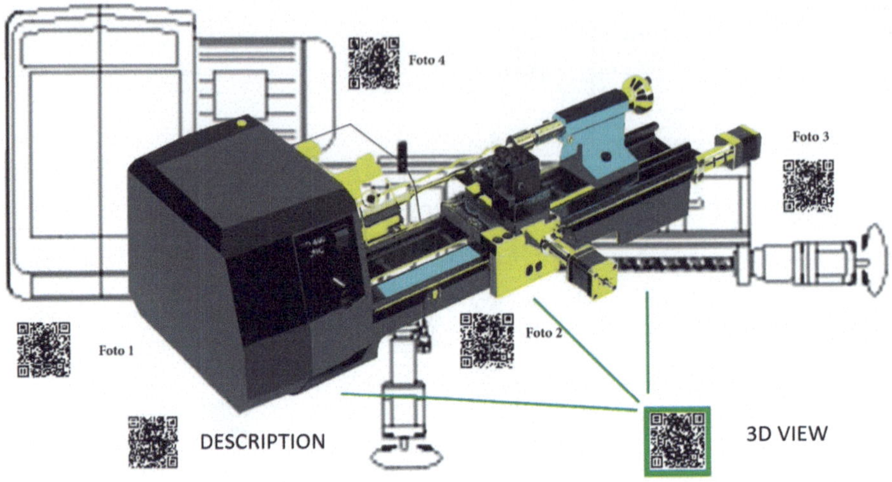

Fig. 7.12 3D model after scanning 3D view markers

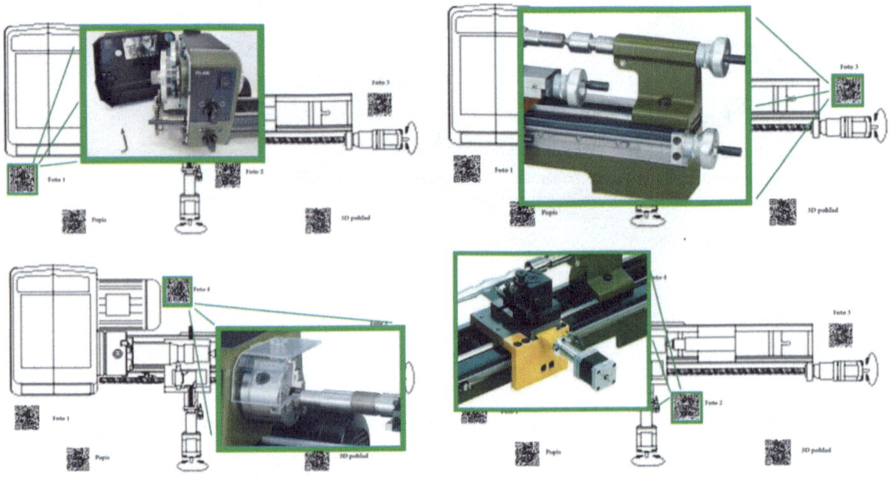

Fig. 7.13 Display descriptive photos to markers

3 Results

In the presented procedure, we tried to show a new trend. The advantage is that Marker AR is the technology that advances and becomes known to the general public. As a new element, we used a specific BuildER PRO tool, which is easy to use and is used to target the markers. It also allows to implement various 2D IRQ codes, which is its additional advantage. We consider the use and loading of black and white images as markers to be one of the main advantages of the given software.

Many authors point to the possibility of reading markers by creating a specific application in the Unity and Vuforia environments. This is advantageous, but the problem is that the applications created in these programs must be programmed and each database is separate for a defined database of markers. The great advantage of the design presented by us is that the application can read several 6 markers simultaneously. Another advantage is the simplicity and intuitiveness of operation. In the presented solution, we wanted to show the possibility of augmented reality as an element for providing information in manufacturing companies but also in the education of students, where it is possible to assign various information to various images (markers).

4 Conclusion

It can be said that augmented reality is basically a new technology that is still advancing. AR is gaining more and more supporters and it is used in various fields, such as industry, healthcare, sales, and it also has a great response in entertainment and mobile applications. Markers are increasingly used. This technology can be used almost everywhere, especially in the interior. It is used, for example, in education, where this technology is also supported in development. The combination of real and virtual worlds helps students to properly understand the issues of the teaching process, when some elements of teaching are simulated by a 3D virtual reality model. The Marker AR is applied, for example, in the automotive industry, whether in vehicle simulation or it is helpful in vehicle repairs and construction. Such an application can also provide information on the vehicle technical data, spare parts, and related procedures. We can find another use in construction, in the analysis of building construction, deployment in a specific reality, in the proposals creation, projects, where the simulation of an already built building and placement in a real environment will help us decide of the suitability of realization or eventually necessary changes in the project. Mobile applications are a very suitable means of applying Marker AR. There it is possible to play a game while shooting a marker, find a link to a web page, view a 3D model of practically everything, watch a video, and also test yourself in various areas. Simply in connection with a mobile device, the Marker AR has a borderless application. Marker AR can also serve us as a kind of GPS navigation in the interior. There is a library, for example, where you need to find your way around. That is why so-called virtual libraries are created, which direct us to the book title we are looking for. But it doesn't have to be just a library. These can be various complexes where one needs to orient oneself, such as shopping malls, aqua lands, large sports centers, state institutions, and others. Augmented reality with the use of markers has perspective in the future and can be used almost everywhere. It is just a matter of idea, then realization itself, and finally application in the specified sphere.

Acknowledgments This work was supported by the Slovak Research and Development Agency under contract No. APVV-19-0590, by the projects VEGA 1/0700/20, KEGA 055TUKE-4/2020 granted by the Ministry of Education, Science, Research and Sport of the Slovak Republic, and by the Project of the Structural Funds of the EU, ITMS code: 26220220103.

References

1. Husár J, Knapciková L (2019) Exploitation of augumented reality in the industry 4.0 concept for the student educational process. In: INTED 2019: The 13th annual International Technology, Education and Development Conference, IATED, pp. 4797–4805.
2. Azuma R (1997) A survey of augmented reality. Presence: Teleoperators and Virtual Environments 6(4):355–385
3. Sobota B, Hrozek F (2015) Virtual reality systems. Košice: TU v Košiciach, pp. 260.
4. Ondrejka P (2013) Design of a guide in an environment of augmented reality. Diploma work,. Masaryk University in Brrno, pp. 83
5. Hugues O, Cieutat J, Guitton P (2011) GIS and augmented reality: state of the art and issues. In: Fuhrt B (ed) Handbook of augmented reality, 1st edn. Springier, New York
6. OpenGL. OpenGL Headline News. OpenGL. https://www.opengl.org/., last accesed 2021/1/24
7. Kascak J et al (2019) Implementation of augmented reality into the training and educational process in order to support spatial perception in technical documentation. In: IEEE 6th International Conference on Industrial Engineering and Applications (ICIEA). IEEE, Tokio, pp 583–587
8. Židek K et al (2020) Digital twin of experimental smart manufacturing assembly system for industry 4.0 concept. Sustainability 12(9):3658
9. Strategy. Strategy-hnonline. https://strategie.hnonline.sk/marketing/777468-ciarovy-kod-oslavuje-60-rokov, last accessed 2021/1/15.
10. QR Code., https://sk.wikipedia.org/wiki/QR_k%C3%B3d Last accessed 2021/1/21
11. Tokorova M et al (2020) Augmented reality as a support tool in machining process. Tem Journal 9(1):407–411
12. Perakovic D et al (2019) Information and communication technologies within industry 4.0 concept. In: Advances in design, simulation and manufacturing, lecture notes in mechanical engineering. Springer
13. Kabelka P (2014) Marker optimization for augmented reality applications Mendel University in Brno. Bachelor Work
14. 3D-print Tip: 6 handy applications with which you can create 3D printing models for free. https://www.3d-tisk.cz/tip-6-sikovnychaplikaci-se-kterymi-muzete-bezplatne-vytvaret-modely-pro-3d-tisk/, last accessed 2021/1/16
15. Lavieri E (2015) Getting Started with Unity 5 [e-book]. Packt Publishing
16. PTC. Innovate With Industrial Augmented Reality. PTC. https://www.ptc.com/en/products/augmented-reality, last accesed 2021/1/16.
17. BuildAR. Human Interface Technology New Zeeland. https://www.hitlabnz.org/index.php/products/buildar/. Last accesed 2021/1/16.
18. Wikipedia. ARTag. Wikipedia, the free encyclopedia. https://en.wikipedia.org/wiki/ARTag, last accesed 2021/1/16.

Chapter 8
Increasing the Client Satisfaction by Exploitation of Customer Relationship Management in Construction Projects

Tomáš Mandičák, Peter Mésároš, Annamária Behúnová, Marcela Spišáková, and Andrea Kanáliková

1 Introduction

The success of construction projects depends on several factors [1]. Many studies address this issue and point to many factors that can potentially affect projects' success [2]. Many studies point to the importance of monitoring key performance indicators (KPIs) [3]. Several authors point to inconsistencies in information when comparing the success of construction projects. Key performance indicators can measure performance. The aim of the construction project is the fact that the client wants several parameters to be met, such as construction time, budget, quality, safety, etc. KPIs are a tool for measuring these parameters using validated methods [4].

Authors of the study were talking about these KPIs: cost, time, productivity, quality, and people [5]. Other authors perceive the boundaries of KPIs as cost, quality, and productivity [6]. They also stated that important KPIs are as well safety and environmental aspects. But Iyer et al. mentioned the importance of customer and client satisfaction [7]. Another study expanded the list of KPIs, there are cost and time; productivity; client satisfaction; quality; people; safety; regular and community satisfaction [8]. KPIs are economic factors of success. This is often a question of costs, revenues, construction time, and client satisfaction with the construction. This is little analyzed and researched. Client satisfaction (product's and service's view) is one of the basic KPIs. Table 8.1 provides an overview of research to identify KPIs of construction projects.

T. Mandičák (✉) · P. Mésároš · A. Behúnová · M. Spišáková · A. Kanáliková
Technical University of Košice, Košice, Slovakia
e-mail: tomas.mandicak@tuek.sk; peter.mesaros@tuke.sk; annamaria.behunova@tuke.sk; marcela.spisakova@tuke.sk; andrea.kanalikova@tuke.sk

L. Knapcikova et al. (eds.), *Sustainable Management of Manufacturing Systems in Industry 4.0*, EAI/Springer Innovations in Communication and Computing,
https://doi.org/10.1007/978-3-030-90462-3_8

Table 8.1 Key performance indicators in construction projects–literature review

Source	Key performance indicators
[9]	Sales revenue Net profit margin Productivity Sales growth Cost of customer acquisition **Customer loyalty and retention** Monthly website traffic
[10]	Operating cash flow Working capital Current ratio Debt to equity ratio LOB revenue vs. target LOB expenses vs. budget Accounts payable turnover Accounts receivable turnover Inventory turnover Productivity Quick Ratio **Customer satisfaction**
[11]	Financial KPIs (profit, dept asset ratio) Marketing KPIs (return on advertising spend, **customer satisfaction**, time to payback) employee (efficiency and productivity, quality)
[12]	Cost Time Defects Productivity Variance Profitability **Client satisfaction**
[13]	Productivity Quality Financial Innovation Employer **Customer (requirements, satisfaction)**
[14]	Cost Time Predictability Defects Productivity Turnover Profits
[15]	Safety in many points of view (e.g., employee health, documentation, responsibility, etc.

Probably one of the largest studies in this area was the building developer's study in Nigeria [15]. This study identified safety as a key performance indicator. According to another study, the critical aspect is above the efficiency indicator. Especially in projects where energy issues are addressed [16].

Many studies consistently point to one important key performance indicator, that is, customer or client satisfaction. The construction industry produces products that have a long life cycle [17 MARCEL]. Client satisfaction is, therefore, critical. This puts pressure on the cavity. The materials used should be of high quality [18]. However, communication and relationship with the client are equally important. It is precisely for communication and relationship management that there are many possibilities available today, which are offered by information and communication technologies.

Customer Relationship Management (CRM) is an indispensable part of enterprise management [19]. This is particularly true in the current knowledge age. Raw data or information currently are not sufficient [20]. Their amount is growing every second. This is the cause of pressure and emphasis on quality and relevant information with added value [21]. CRM systems are used for the process of obtaining, collecting, storing information, and analyzing too. A good relationship between customers and enterprise presents an essential part of the objectives of every enterprise. This objective will also participate in the main objective of the enterprise, thus making a profit. This objective also helps other partial objectives and is as follows: increasing the enterprise's value and ensuring the solvency of the enterprise. CRM systems are a tool to gain loyal customers.

Customer relationship management is a strategy whose main objective is to understand customers better. Their needs should be fulfilled efficiently to achieve maximum satisfaction. Information about the customer is collected through CRM systems [19]. This information is an intermediary to achieve the most significant degree of satisfaction. With the increasing focus on e-business applications, it has become a significant component of the right CRM e-business strategies [22].

The CRM system contains several sub-modules. Figure 8.1 describes in more detail the possibilities of CRM systems for improving relationships and communication with clients.

The basic functionalities include sales calls, e-mail marketing, documents management, projects documentation, e-mail integration, and relationship management [23]. Important part of CRM systems is creating analysis and communication (Fig. 8.2). It includes order, promotion, and client interaction. It is for increasing customer or client loyalty [24].

Construction project management is a demanding process, and it is very important to define the relationships that arise between individual participants in construction projects as well as clients. Just defining who the customer or client is in these relationships is needed for this research. Figure 8.3 specifies the relationships that enter into the management of construction projects and also specifies the status of the customer and the client.

The first relationship is defined between the investor and the contractor. The investor is the one who pays for the service and the contractor, the one who offers. Their information flows mostly lead to the contractor. The contractor is in the client's position. He keeps records of the investor. The second relationship that arises here is the investor–designer relationship. The designer is in the role of a client. Another relationship is between the contractor and the subcontractor. The

Fig. 8.1 CRM modules and functionality [23]

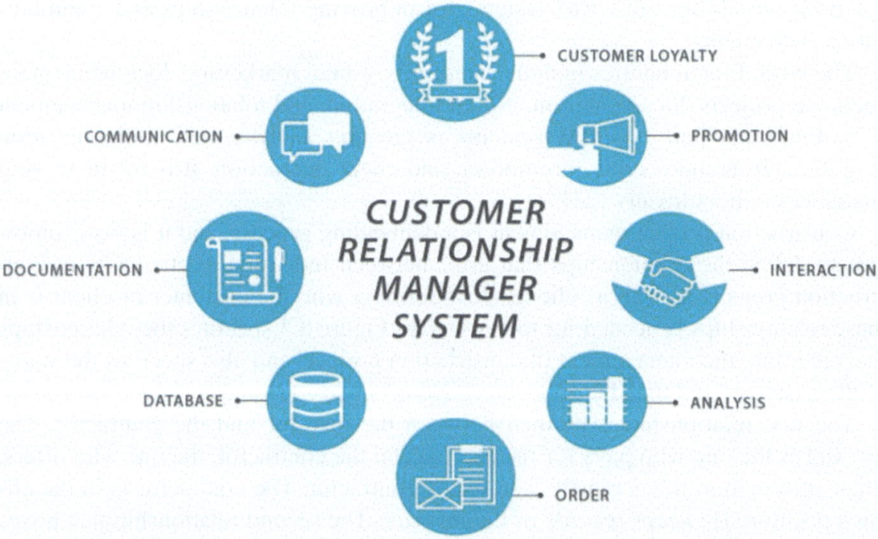

Fig. 8.2 CRM systems purposes [24]

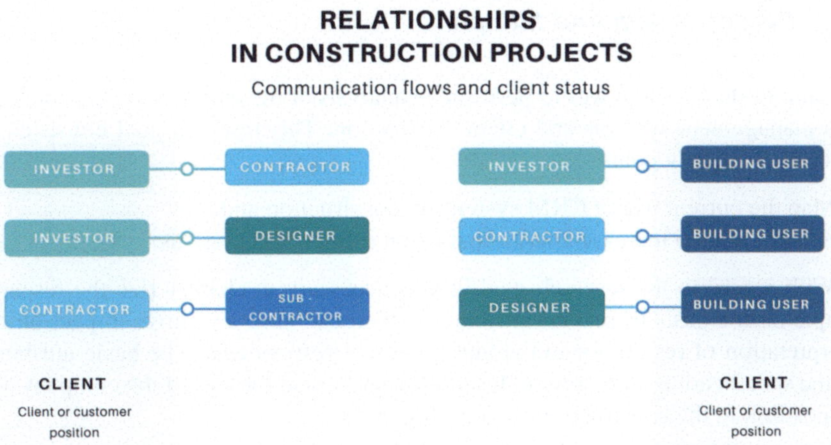

Fig. 8.3 Relationships in construction projects (research purposes)

contractor is in the position of the client in this relationship. Other relationships are defined to the building user, who is a client in these cases. These are relationships between: investor and user of the building; contractor and the user of the building; designer and the user of the building.

From these relationships, it is clear that the CRM system can be used not by clients but by service providers. It is possible to specify the following groups: investor, contractor, subcontractor, and designer. It is these groups that also represent the research sample.

The client's view and satisfaction are also critical. However, these must be done by the service provider himself.

Based on the above, several research questions arise here, which are not much researched in the field of construction:

- What is the level of use of CRM systems for customer relationship management?
- What are the differences between the participants in construction projects?

2 Methodology

Research aimed at using CRM systems and increasing client satisfaction is based on the use of primary and secondary data. The following sections describe the data collection as well as the clarification of the processing method. As well as the goal of the research and the research sample.

2.1 Research Aim and Sample

The aim of the research was to perform an analysis of the use of customer relationship management systems and client satisfaction. This research goal consisted of fulfilling the purity goals:

- Map the current use of CRM systems in construction and
- Specify how to increase client satisfaction using CRM systems.

With regard to the research goal, it was necessary to characterize the research sample on the basis of several attributes, which can have a positive impact on the interpretation of results for individual groups of respondents. The basic attributes for the specification of the research sample represented the size of the company and the position in the construction project (Fig. 8.4).

Large companies participated in the representation of almost 13% of respondents, which is a very positive signal in terms of the total number of large construction companies operating in the construction market. The SMEs were evenly represented. In this case, it is important to take into account the opinion of microenterprises operating in the market, and in many cases, there are a huge number of them, which are also an important unit in the economy. It was these companies that were also very willing to answer questions about the topic under study.

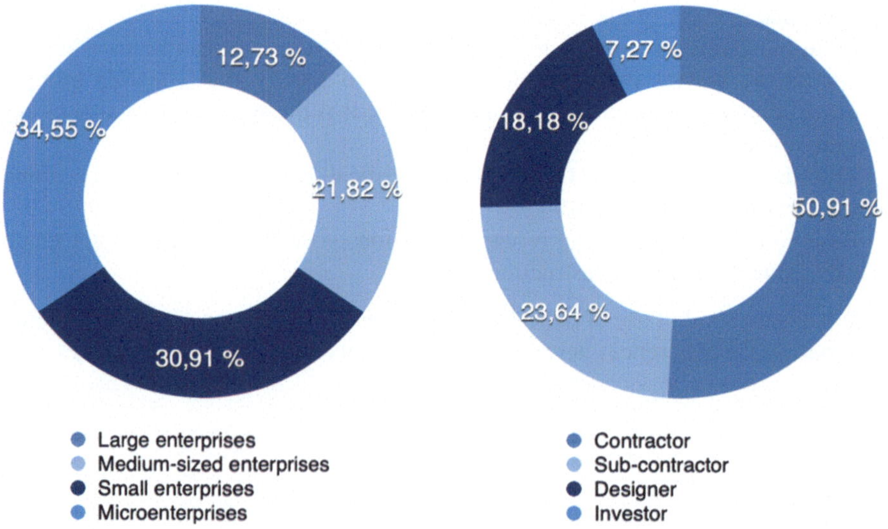

Fig. 8.4 Research sample by enterprise size and participant of construction project

2.2 Data Collection

Several types of data were used in the research. Secondary data were used to determine the research problem and map the situation abroad. These were mostly from valuable scientific databases. Also, for the determination of research questions, inquiries and short interviews were conducted with a small research sample, which had the task of defining the suitability of the questions and also providing feedback on the need to implement this research issue.

For the purposes of obtaining and processing the research results, primary data were used by collecting through a questionnaire survey and interviews. This means that both quantitative and qualitative methods have been used in research in order to more accurately analyze and clarify certain scientific conclusions. The description of the use and acquisition of data is described in Table 8.2.

As far as the identification of research questions is concerned, these have been selected on the basis of the above literature and after review of previous research that has addressed similar issues. These research questions were then consulted with practitioners. They were experienced project managers and financial managers in the industry. Each research group was represented by an expert, meaning that it went from manager to investor, contractor, subcontractor, and designer.

An online questionnaire was used for the main research conclusions. Questions approved by practitioners were distributed online to respondents. The invitation to participate in the survey also included information for respondents, which included an explanation of the probability and instructions on how to understand and answer individual questions.

The questionnaire was divided into several thematic sections. The first part contained questions focused on the characteristics of the respondent. These were information such as the size of the company, the position of the respondent, the experience of the respondent, the participant in the construction project, the use of foreign capital, etc.

The second part dealt with questions focused on the analysis of the current state of use of CRM systems. As well as information on how long they are used and how many people work with them. The third part examined the impacts and the increasing experiences of client satisfaction. These questions were designed so that the respondent clearly and measurably identifies the impact of the CRM system on client satisfaction.

This section also includes information on whether the company has also verified client satisfaction and whether it was relevant to assessing the impact of using CRM systems. The aim was to guide the respondent to the most accurate answer possible, based on which it is possible to quantify the impact.

Table 8.2 Utilization of data collection and processing methods for research purposes

Research purpose	Data type	Data collection methods
Problem statement	**Secondary data**	Scientific databases and previous researches, Interview
Research activities	**Primary data**	Questionnaire, Interview

2.3 *Data Processing*

Data processing was based mainly on statistical methods designed for quantitative research. Respondents had to answer questions on a scale from 1 to 5 (1—low utilization rate and 5—high utilization rate). Also, for questions focused on the impact and consideration of client satisfaction (1—low impact on client satisfaction, 5—high impact on client satisfaction). The suitability of the research questions and answers was also tested with Cronbach's alpha.

These data were then collected and analyzed through statistical and spreadsheet programs. The measured values were averaged and the number of responses was generated. The Kruskal–Wallis test was used to determine the significance of the conclusions.

He had to confirm or refute the research conclusions and claims at the 95% significance level.

These quantified results were subsequently discussed, and conclusions were reached. In this step, connections were also sought as to why certain phenomena or findings occurred. Qualitative research was also used in this part, when selected conclusions were consulted with experts and their views from practice.

3 Results and Discussion

The use of information and communication technologies depends on many factors. Based on the interviews conducted, several factors influencing the implementation of ICT in general were identified. These factors also apply to the implementation of CRM systems. These are, above all, high investment costs, operating costs, complexity of the system, the need to employ IT specialists, the need to employ new administrative staff, reluctance to accept changes and innovations, and concerns about the failure of system implementation.

On the contrary, increasing sales, customer satisfaction, facilitating work, increasing productivity as well as saving time can be included among the main motivators for the implementation of CRM systems. Monitoring customer or client satisfaction is very important, which is also one of the key performance indicators.

The first monitored indicator was the use of CRM systems in the monitored sample. It must be said at the outset that the rate of use of CRM systems in the construction industry is relatively low (Fig. 8.5). Companies have little and no implementation of CRM systems for customer relationship management.

As can be seen in Fig. 8.5, large enterprises use CRM systems to the greatest extent. However, this value reaches such a low number and only 2.35, which is not really a high value. The utilization rate is decreasing toward the minimum, even according to the size of the enterprises. CRM systems are almost never used by micro-enterprises. This finding is not surprising and is identical to research expectations.

The assumption that the use of CRM systems also depends on the size of the company is still relevant and at least this trend has been captured by research.

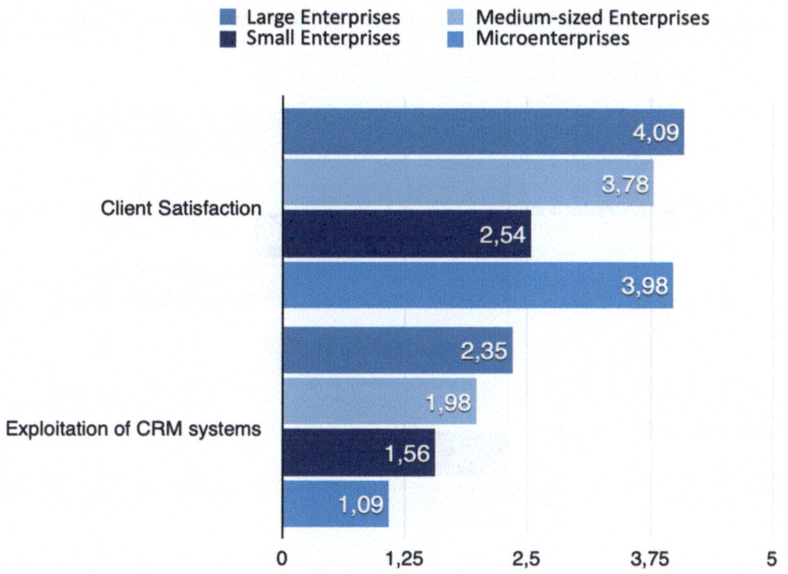

Fig. 8.5 Exploitation of CRM systems and client satisfaction by enterprise size

However, what is more important for the fulfillment of the main research goal is the fact that the ponies that use this tool evaluated that their clients are more satisfied. However, the significance of this statement should also be demonstrated by a statistical test.

However, this applies to all research groups. Quite an unexpected trend has happened in microenterprises. Their customers show a relatively high level of satisfaction, which is at the expense of customer satisfaction of large companies. However, this refutes the assumption that the size of the company has an impact on this and the trend is associated with the size of the company and a higher level of customer satisfaction.

This networking is important to discuss. Based on the expert testimony, one estimate was made as to why this is so. This may be a situation where microenterprises strive to be more personal and willing to do as much as possible for the customer. Client satisfaction can therefore be high not because the company has set up processes and builds a relationship with the customer through information and communication technologies, but because it has a personal approach to the customer. The customer does not perceive whether the supplier contacts him through the CRM system or uses common and elementary technologies. He is also not interested in whether he has written information about his requirements in the system or just on paper. The customer perceives the fact that his request has been met and thus shows a high degree of satisfaction.

Another view is given by the analysis of the perception of client satisfaction in the use of CRM systems according to the division of the research sample into the participant in the construction project (Fig. 8.6).

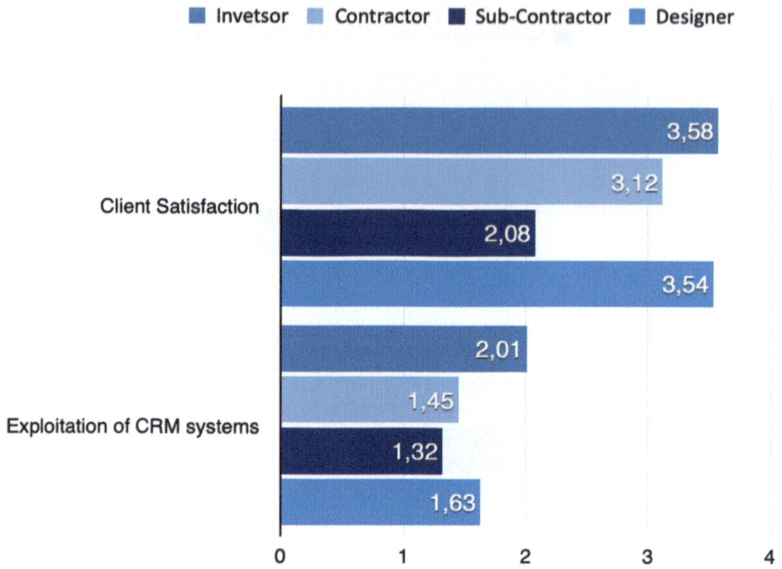

Fig. 8.6 Exploitation of CRM systems and client satisfaction by participant of construction project

Utilization rates range from 1.32 for subcontractors to 2.01 for investors. Here, however, the trend is clearer, namely that for groups that use CRM systems to a greater extent, customers are more satisfied.

The investor-building user relationship is specific, in that the client is a building user, but in reality, not all services are done by the investor (developer). It should be noted here that this research abstracts that the investor is also a building user. Nevertheless, the customer (building user) evaluates the investor (developer) and he does not care who actually built the building. His satisfaction is linked to the communication of the investor (developer).

On the contrary, the investor–contractor relationship is different. The satisfaction of the investor as a client depends on how the contractor communicates with him and what design of the product he will provide. Again, it should be emphasized that construction is not retail. Expecting use and conditions similar are inadequate. Therefore, this behavior is also specific.

The analysis of the impact of CRM systems and the growth of client satisfaction brought only a comparison of customer satisfaction before and after implementation (Fig. 8.7).

Figure 8.7 shows that after the implementation, the number of clients increased and also their satisfaction. The longer the CRM system was implemented in the company, the higher the customer satisfaction. It turned out that there is a trend that indicates that after the implementation of the system, customer and customer satisfaction also increased. Companies that did not use the CRM system for the reference period (i.e., at least 2 years) were excluded from this statistics.

On the contrary, the rate of time spent with administration per client decreased after the implementation of the CRM system (Fig. 8.8). This means that the CRM

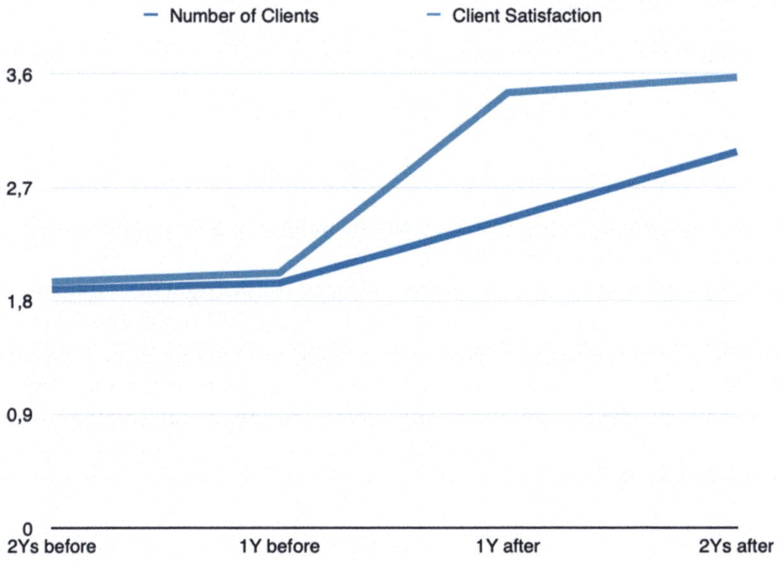

Fig. 8.7 Comparison of number of clients and client satisfaction before and after implementation of CRM system

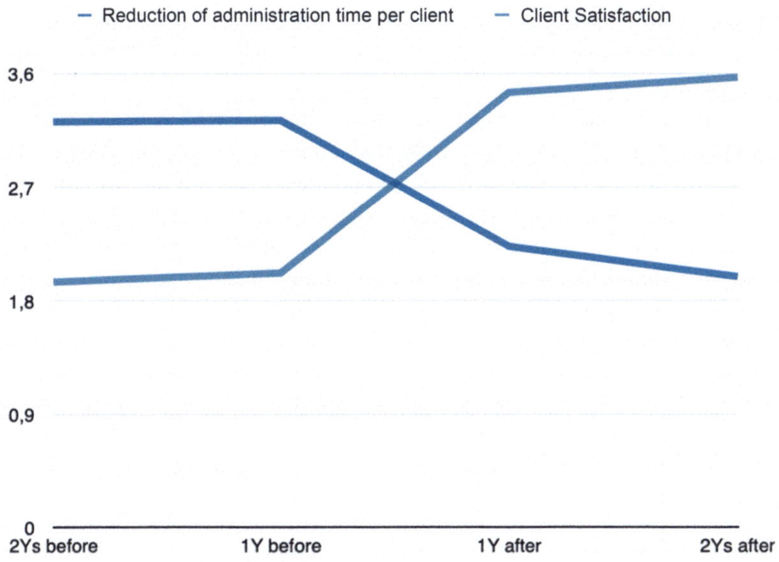

Fig. 8.8 Comparison of reduction of administration time per client and client satisfaction before and after implementation of CRM system

Table 8.3 Kruskal–Wallis test

Exploitation of CRM	Impact rate of CS	p
Yes	**3.34**	0.0736
No	**2.57**	

system saves employees' time and they can be more productive. This has also been mentioned as another important benefit.

The crossover on the graph shows that the trend breaks right after the implementation of the CRM system. This is another signal that CRM represents benefits.

To confirm the impact of the CRM system on the growth of customer satisfaction in construction, statistical testing was performed by Kruskal–Wallis test (Table 8.3).

The P value reached 0.736, which represents a relatively high statistical significance, but since the value determined by us was 0.05, this statistical test did not confirm these results.

4 Conclusion

Research focused on the analysis of the use of CRM systems and the increase in client satisfaction in the field of construction has pointed to several trends. Overall, the use of information and communication technologies in construction has not yet reached its potential. The reason is often factors that hinder the implementation of information and communication technologies to a greater extent. The same goes for CRM in systems.

Among the main barriers to the implementation of CRM systems are the high investment costs for CRM systems as well as concerns about operating costs. It is also a perception of the reluctance to accept change and accept innovation, the effort to learn new things, and also the concern of managers to accept a new workforce. All these rather play a negative role in the implementation of new CRM systems.

On the other hand, the main motivators include an increase in orders, positive expectations from an increase in employee productivity, and also an increase in customer satisfaction.

The research pointed to the trend that the implementation leads to an increase in customer satisfaction in the construction sector. However, at a statistical significance of 0.05, this was not confirmed.

Based on the above information and considerations, the results to date can be summarized as follows:

- The use of CRM systems in construction is relatively low.
- CRM systems are being used more by larger companies.
- Companies that use CRM systems report a higher level of customer satisfaction.

- The level of client satisfaction after the introduction of the CRM system grows, the longer the system is implemented.
- The level of administration time falls by implementation of CRM.

Client satisfaction is one of the key performance indicators in the construction industry. However, client satisfaction is a very sensitive topic and there are many factors that influence it. However, CRM systems have been proven to help companies achieve this goal. CRM systems also have other benefits, which this research also mentioned.

Acknowledgments This work was supported by the Slovak Research and Development Agency under contract no. APVV-17-0549. Paper presents partial research results of project VEGA 1/0828/17 "Research and application of knowledge-based systems for modelling cost and economic parameters in Building Information Modelling." The paper presents partial research results of project KEGA 059TUKE-4/2019 "M-learning tool for intelligent modeling of building site parameters in a mixed reality environment."

References

1. Radziszewska-Zielina E, Glén M (2016) The application of the electre I method for the selection of the floor solution variant, Advances and Trends in Engineering Sciences and Technologies – Proceedings of the International Conference on Engineering Sciences and Technologies. ESaT 2015:359–365
2. Egbu CO, Botterill K (2002) Information technologies for knowledge management: their usage and effectiveness. ITcon – ICT for Knowledge Management in Construction, pp 125–137
3. Kršák B, Kyšeľa K (2016) The use of social media and internet data-minig for the tourist industry. J Tour & Hosp 5:1–3., ISSN 2167-0269
4. Wang WC, Weng SW, Wang SH, Chen CY (2014) Integrating building information models with construction process simulations for project scheduling support. Autom Constr 37:68–80
5. Adnan E, Khalid A-H, Sherif M (2006) Causes of contractor's business failure in developing countries: The case of Palestine. J Constr Dev Ctries 11(2):1–14
6. Ugwu OO, Haupt TC (2007) Key performance indicators and assessment methods for infrastructure sustainability – a South African construction industry perspective. Build Environ 42:665–680
7. Iyer KC, Jha KN (2005) Factors affecting cost performance: evidence from Indian construction projects. Int J Proj Manag 23:283–295
8. Samson M, Lema NM (2002) Development of construction contractors performance measurement framework, 1st International Conference of Creating a Sustainable
9. Karslon K (2019) 12 Business Metric That Every Company Should Know. [Online] https://www.scoro.com/blog/12-business-metrics, last accessed 2019/11/03.
10. Gerber B (2018) 12 Key Financial Performance Indicators You Should Be Tracking. [Online] https://www.accountingdepartment.com/blog/12-key-performance-indicators-you-should-be-tracking, last accessed 2019/11/03.
11. Wiliams A (2017) 11 Key Business Performance Metrics for Metrics Better Operations. [Online] https://alistemarketing.com/blog/business-performance-metrics, last accessed 2019/111/03
12. Constructing Excellence (2009, May). UK Construction Industry KPIs. Retrieved May 16, 2009, from Constructing Excellence Website: www.constructingexcellence.org.

13. Sousa SD, Aspinall EM, Rodrigues G (2006) Performance measures in English small and medium enterprises. Benchmark Int J 13(1/2):120–134
14. Kaplan R, Norton D (1992) The balanced scorecard – measures that drive performance. Harvard Business Review Vol 70, March–April.
15. Mahmoud AS, Ahmad MH, Yatim YM, Dodo YA (2020) Key performance indicators (KPIs) to promote building developers safety performance in the construction industry. *J Ind Eng Manag* 13(2):371–401. https://doi.org/10.3926/jiem.3099
16. Walker NL, Williams AP, Styles D (2020) Key performance indicators to explain energy and economic efficiency across water utilities and identifying suitable proxies. J Environ Manage 169:1–10
17. Rosová A, et al. (2020) Case study: the simulation modeling to improve the efficiency and performance of production process. Wireless networks: the journal of mobile communication, computation and information, pp. 1–10
18. Knapčíková L, et al. (2018) Advanced Materials based on the Recycled Polyvinyl Butyral (PVB). 2018. MMS Conference 2017. – Ghent: EAI, 2018, pp. 1–9
19. Čarnický Š, Mesároš P (2013) Business Intelligence and Knowledge Management: A Business Perspective. – Brussels: EuroScientia vzw, 2013. 168 pp. ISBN 978-90-818529-1-3.
20. Čarnický Š, Mesároš P (2009) Informačné systémy podnikov. Bratislava: Ekonóm. 233. p. ISBN 978-80-225-2676
21. Kršák B, Tobisová A, Sehnálková M (2011) Information technologies and their using during firm's financial health modelling. Trans Univ Košice 4:35–40. ISSN 1335-2334
22. Madu CN, Kuei C (2005) ERP and supply chain management. Faifield: Chi Publishers. 218 s. ISBN 0-9676023-4-3
23. https://www.perfectviewcrm.com/what-is-crm/
24. http://lsinet.co.uk/services/crm/

Chapter 9
Use of Probabilistic Risk Assessment Methodology for Providers of Services in a Virtual Environment

Petra Zorić, Mario Musa, and Tibor Mijo Kuljanić

1 Introduction

In high technology, organizations have become too dependent on information systems, while various forms of information and data kept on numerous media have become the most valuable resource of business systems. In that context, business system reputation has also become a form of an information asset. On the other hand, there are many threats to such property. The number of threats and their manifestations is increasing on an everyday basis. For that very reason, many organizations identify their information as significant assets that they need to protect by internal security checks.

Information security must be taken care of because companies' value is mostly concentrated in the value of information. To fully understand and be able to protect their information, organizations must be acquainted with the concept of information security. It is one of the key parameters which influence organizational exposure to business risk. Control of information starts with the assessment of its risk exposure. The world has been growing and technologically improving for years without questioning the security in the way it is questioned now. This resulted in cyber threats and vulnerability deeply integrated into the architecture of information systems and

P. Zorić (✉)
University of Zagreb, Faculty of Transport and Traffic Sciences, Zagreb, Croatia
e-mail: petra.zoric@fpz.unizg.hr

M. Musa
Hrvatska Lutrija d.o.o., Zagreb, Croatia
e-mail: mario.musa@lutrija.hr

T. M. Kuljanić
HEP ODS d.o.o., Zagreb, Croatia

L. Knapcikova et al. (eds.), *Sustainable Management of Manufacturing Systems in Industry 4.0*, EAI/Springer Innovations in Communication and Computing,
https://doi.org/10.1007/978-3-030-90462-3_9

systems which support their correct functioning. As a result, a cyberattack represents a level of challenge the organizations are only preparing for.

Risk analysis of undesirable and harmful situations is one of the phases of the safety life cycle. The assessment of risk or adverse events affects the adoption of different measures based on the frequency of events and the consequences of the event's severity. Scenario development to predict and prepare for causal accidents and extraordinary events in the nuclear, chemical, and energy sectors has been used to prioritize and quantify and undertake actions. Probabilistic Risk Assessment (PRA) is one of the methodologies based on scenario development elaborated for the needs of high-risk and complex systems in nuclear power plants. This methodology allows for quantification of the probability of events related to various types of accidents and quantifies the occurrence of various incidents.

This chapter presents a use case of information security risk assessment using fault tree and event tree methods as part of PRA methodology. The operational concept is based on the availability of service providers' data essential for service delivery in the virtual environment of games of chance. The chapter also presents these methods in a complex information system such as that of the company Hrvatska Lutrija.

2 Literature Research

The development of information and communication systems leads to the possibility of reducing data security in organizations. The gaming industry in a virtual environment for the sake of scale and finance is interesting to study. A reliable and stable ecosystem for delivering virtual services, i.e., virtual games, is essential. The Industry 4.0 paradigm can also be applied in the gaming business [1].

Great efforts are currently being made to maximize the protection of information security. One of the ways of protection is the methods of information security risk assessment in organizations. For this reason, this chapter presents the application of the fault tree and event tree methods to the example of a real information system.

Information security is a state of confidentiality, availability, and integrity of data [2]. The CIA triad, i.e., confidentiality, availability, and data integrity, are the primary information security goals. New technologies currently available, such as artificial intelligence or machine learning, can focus on all the triad features [3]. The CIA triad is crucial in games of chance, given that user data in the specified environment must be confidential and accessible, while the system must be 100% reliable. The focus of the environment is on the security of the data and information systems involved in their processing.

The establishment of modern business environments is crucial for the very security of information exchanged in such an environment [4]. The availability of services in the gambling environment is one of the essential elements of security of such a system and plays a vital role in its operation and the work of the services provided to end users through the system [5, 6].

Identifying and assessing security risks and adopting appropriate risk mitigation strategies is one of the main challenges of complex information and communication systems [7]. Security risk considers the probability and impact of the exploitation of vulnerabilities of the organization's information assets, which causes damage to the organization [8]. Information assets are subject to several threats that arise due to omissions, intentional and unintentional human error, technical reasons, regulatory noncompliance, policies and procedures, and force majeure [9]. Mitigation of these threats has become an essential organizational strategy that uses principles and practices based on prevention and response procedures [8].

The risk management system establishes a mechanism for controlling and managing the acceptable level of risk to the organization. It consists of identifying and classification of all information assets, analysis of all possible risks, risk level assessment, damage identification, and determination of risk management methods and controls that will minimize the risk of system failure [10]. Risk assessment is often performed in a minimum of two iterations. If sufficient risk assessment information is not provided in this way, detailed analyses should be performed using other risk assessment methods [11].

When the need for a more detailed asset-specific analysis or critical business process arises, more complex methodologies can incorporate the complexity of the relationships among information assets. Recently, many research groups have conducted PRA in various technological fields, which have led to findings in the assessment and mitigation of risks inherent in a particular field of research [12–17].

There are numerous examples of modern data centers equipped with IoT sensor systems for monitoring and controlling air conditioning systems. The application of IoT raises additional security issues, and it is necessary to pay special attention to Distributed Denial of Service (DDoS) attacks on the system itself [18–20].

3 PRA Methodology for Information Security Risk Assessment

PRA is one of the methodologies applicable for a more detailed risk assessment analysis associated with complex technical systems. The framework of the methodology is a combination of a deductive and inductive logic model and was presented more than 30 years ago in the *WASH-1400 Reactor Safety Study* in the nuclear industry [21].

Although it was initially used in the nuclear industry as a methodology for calculating nuclear power plant risk assessment, PRA is also used in information security risk assessment. The more frequent PRA application is due to increasingly complex information and communication systems that have more and more interdependent parameters. A particular hazard can be assessed by the convolution of system fragility and hazard curves expressing the probability of exceeding as a function of the intensity measure used to assess the hazard [22].

In PRA, the risk is defined as the possible harmful consequence of a particular activity. It is characterized by two components [23]:

- Scope/strength/impact of possible adverse effects and
- The probability of any consequence.

There are three fundamental questions that the PRA most often answers [24]:

1. What are the initial events that lead to adverse consequences?
2. What are and how severe are the damages to which the subject may be exposed due to the initial event's specific occurrence?
3. What is the probability of occurrence of an adverse consequence, i.e., what is the frequency of its occurrence?

The methods used to answer the previous questions are Event Tree Analysis (ETA) and Fault Tree Analysis (FTA). The traditional PRA methodology uses ETA to list possible scenarios, while FTA is used to find the causes of individual scenarios. These methods may include simultaneous failures of technical components and failures caused by human error, which allows a broader analysis of the causal factors that lead to the final event (accident or technical failure) [25].

By creating different scenarios, it is possible to conduct both quantitative and qualitative risk assessments. In general, terms such as start event, transition event, and event tree are used to create these scenarios. Consequences are expressed numerically, while probabilities are expressed as percentages or frequencies. The absolute risk is the expected loss expressed as the sum of the products of the consequences multiplied by the corresponding probabilities. PRA can be performed for two types of events: initial internal events and initial external events. Internal initial events can be defined as a system failure caused by an operator error, while initial external events are associated with the occurrence of natural phenomena such as floods or fires.

ETA is an analytical technique used to assess the reliability of a system generally. It is used to assess the course of a process and events that can lead to an accident and uses the tree's visual structure to show the logic and development of such events. Its goal is to determine whether there is a possibility that the initial event may develop into a series of failures or be sufficiently controlled by the security systems and procedures implemented in the system design. ETA can result in multiple different outcomes from a single initial event and allows probabilities to be obtained for each outcome. Different outcomes represent different branches of the event tree. Based on this analysis, it is possible to report equipment failures or faults and identify improvements in the security systems and other security functions of an information system [26].

ETA is an inductive procedure for identifying and assessing all consequences in the analyzed technical system that may occur after the initial event, considering additional events and factors and the functioning of safety barriers. A primary starting point is an unwanted event that can have a detrimental effect on the information system. For each initial event, additional conditions are identified that could contribute to the realization of undesirable consequences. Obstacles created to address the initial event are then identified. The consequences and probabilities of each scenario are estimated for each initial event. The set of probabilities of all scenarios makes up the risk profile of the technical system. The aim is to assess measures such as safety features that are effective in reducing side effects. The ETA method can be used as a complementary FTA method to determine an adverse event's consequences.

ETA and FTA use the same mathematical and logical techniques. However, the ETA considers the impact of a failure of a particular component in the system and determines the effect of such a failure on the overall system risk or reliability. On the other hand, the FTA analyzes how an unwanted main event occurs. When the risk exposure is too high, i.e., when the consequence's trajectory is not acceptable, the development of strategies that can apply the risk exposure is approached.

FTA is a method of determining an event's probability that may lead to an adverse event. Events are determined deductively, i.e., based on the peak of the event and how the causes that could have caused an undesirable event are sought. It can be used for all types of system-level risk assessment processes. The purpose of this analysis is to effectively identify system failure patterns and mitigate risk before a failure occurs. It is ideal for complex systems because it can visually display a logical way to identify problems, which leads to efficiency.

The analysis is based on Boolean logic and consists of determining the undesirable consequence and the primary events that contribute to the foreseeable undesirable peak event, which leads to the assessment of the risk itself. Each event tree consists of three components: a process diagram, events that have occurred and to which the specified diagram applies, and gates or links between events. The primary events are connected using standard logical operators (AND, OR, etc.) via gates. Events that must exist or occur in conjunction with another underlying event to cause an undesirable event are described using AND connectivity. Primary events that can individually cause an undesirable consequence are described using OR connectivity [22].

The focus is usually on a significant failure, or catastrophic event called the above event and appears at the top of the fault tree diagram. The qualitative analysis consists of determining the different combinations of events that will cause the main event to occur, while quantitative analysis occurs afterward to assess an important event's probability of occurrence [27]. Quantitative analysis allows the calculation of the probability of a peak event occurrence based on partial event values.

As an example of ETA and FTA's use in risk assessment, this chapter will consider the case of failure of the air conditioning system of the hall system in which necessary computer and network devices of the information system of the Croatian Lottery are located. The Croatian Lottery has systems that require high availability, and an increase in the temperature in the hall system (over 40°C) can cause the hardware components to stop working (most equipment is declared to operate smoothly only at temperatures below 40°C). The temperature rise in the event of a failure of the air conditioning system in the hall system quickly occurs due to many computer and network resources that heat the space. In the event of a failure of the primary air conditioning system unit, the redundant unit must be manually switched.

For the air conditioning system to operate continuously, an additional redundant unit, a temperature sensor, and a human response must function.

The traditional risk analysis method does not allow the combination of all elements to determine which errors cause downtime. Therefore, the ETA method will identify the sequence of events that can lead to the air conditioning system's failure and, based on the statistics of previous events, assess the probabilities of these events. The FTA method will analyze each event from the event tree. As a result, a list of the causes of the air conditioning system's failure and the probability that this error on a particular subsystem affects the operation of the entire system will be displayed.

4 ETA Method in the Case of Air Conditioning System Failure

In this use case, the ETA method will be demonstrated through the following four steps: initiating event identification, identification of security checks established in response to the possible occurrence of the initiating event, event tree construction, and evaluation, description, and interpretation of the obtained incident sequences.

The initiating event considered in this use case is the failure of the system room air conditioning system. The system room from this example is used for crucial hardware infrastructure (servers, various network equipment, and data storage system) of the information system of Hrvatska Lutrija. The equipment kept in this room require humidity and temperature maintenance according to the guidelines prescribed by single producers. The organization satisfies these requirements through a separately constructed and implemented air conditioning system. The system comprises two air conditioning units: the primary and the secondary. Only one of them is active, while the other is used as support, and it is redundant. The air conditioning system's functioning is monitored using a sensor for measuring system room temperature and humidity. In the event of temperature or humidity rise, human intervention is needed to regulate the settings of the air conditioning system and/or turning the redundant unit on.

In the event tree (Fig. 9.1), the upward branching represents a positive outcome, while the downward branching represents a negative outcome of the event.

For managing, surveillance, and protection of air conditioning system functioning, a sensor system for measuring temperature and humidity has been implemented, as well as a system of email reporting of nonregulated values read from the sensor, 24/7 monitoring by technical staff, and the possibility of introducing a substitute air conditioning unit, if needed.

The events which can be examined in the system implemented and controlled in this manner are:

1. Temperature sensor reports temperature increase in the system room which exceeds the permitted limit

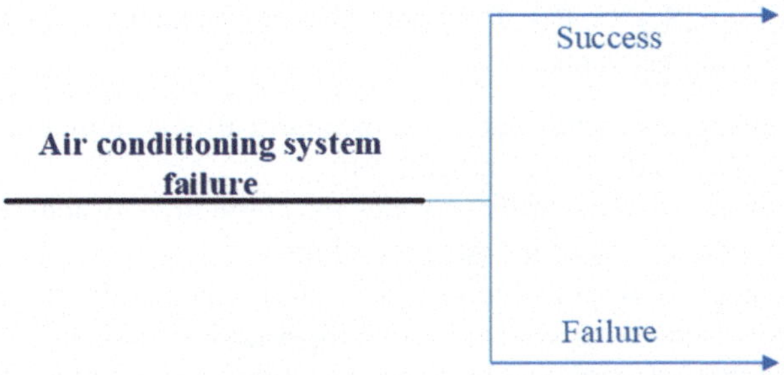

Fig. 9.1 Initiating event

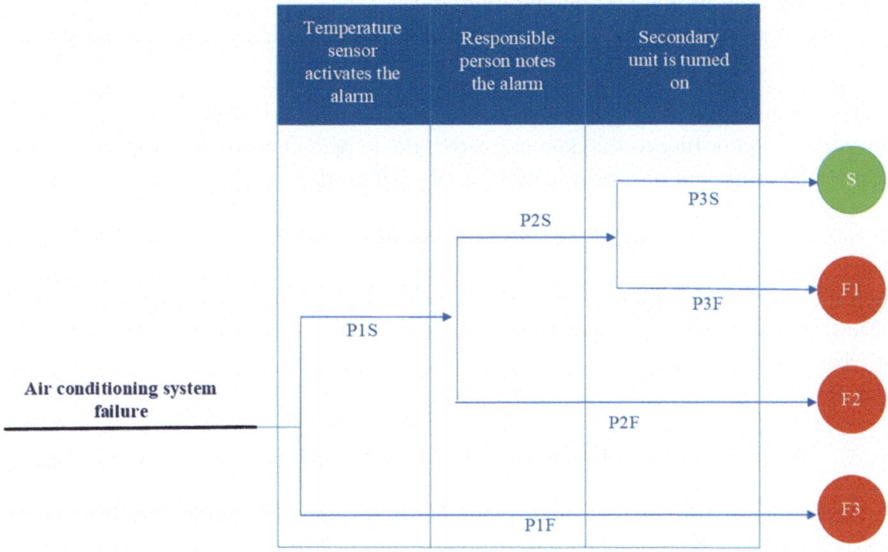

Fig. 9.2 Air conditioning system failure—event tree

2. A member of responsible technical staff notice the alarm activated by the sensor
3. A member of responsible technical staff turns on the redundant unit.

The data registered by now, identified by monitoring the incidents that occurred to the air conditioning system, indicate the following:

- Air conditioning system failure occurs on average twice a year;
- The expected accuracy of alarm activation by the temperature sensor is 99.9%;
- A member of responsible technical staff perceives the alarm in 90% of cases; and
- A member of responsible technical staff successfully turns on the redundant unit in 95% of cases.

The event tree for the described system is presented in Fig. 9.2.

The marks P1S, P2S, P3S, P1F, P2F, and P3F of the event tree in Fig. 9.2 represent the probability of single events:

- P1S—The sensor *successfully* activates the alarm after system room temperature increase which exceeds the permitted limit;
- P1F—The sensor *unsuccessfully* activates the alarm after system room temperature increase which exceeds the permitted limit;
- P2S—A member of responsible technical staff *successfully* notices the alarm activated by the sensor for system room temperature monitoring;
- P2F—A member of responsible technical staff *unsuccessfully* perceives the alarm activated by the sensor for system room temperature monitoring;
- P3S—A member of responsible technical staff *successfully* turns on the system room air conditioning redundant unit; and
- P3F—A responsible technical staff member *unsuccessfully* turns on the system room air conditioning redundant unit.

The probability of single marks of the events can vary depending on the presumptions. According to the data collected, the expected accuracy of alarm activation by the temperature sensor is 99.9%, i.e., 0.999. If

$$\mathrm{P1F} = 1 - \mathrm{P1S} = 1 - 0{,}999 = 0{,}001, \tag{9.1}$$

then the event is complementary to P1S event. According to the data collected, if P2S=0,9, a member of responsible technical staff will notice the alarm in 90% of cases.

P2F event is complementary to P2S event because

$$\mathrm{P2F} = 1 - \mathrm{P2F} = 1 - 0{,}9 = 0{,}1. \tag{9.2}$$

According to the data collected, a member of responsible technical staff will successfully turn on the redundant unit in 95% of cases. P3F event is complementary to P3S event because

$$\mathrm{P3F} = 1 - \mathrm{P3S} = 1 - 0{,}95 = 0{,}05. \tag{9.3}$$

In the same figure, possible outcomes marked with F1, F2, and F3 have the following meanings:

- S—After the failure of the air conditioning system, system room temperature monitoring sensor registered an increase in temperature, which exceeded the permitted limit and successfully activated the alarm, which a member of responsible technical staff successfully noticed and successfully turned on the secondary air conditioning unit of the system room, which resulted in the fall of temperature to the permitted level;
- F1—System room temperature monitoring sensor did not register the increase in temperature exceeding the permitted limit, and the temperature in the room rose uncontrollably until the automatic shutdown of the installed equipment and its probable damage;

- F2—System room temperature monitoring sensor registered temperature increase exceeding the permitted limit and successfully activated the alarm, but a member of responsible technical staff did not notice it, which lead to the same consequences as in the F1 case;
- F3—System room temperature monitoring sensor registered temperature increase exceeding the permitted limit and successfully activated the alarm, which a member of responsible technical staff noticed but did not manage to activate the redundant unit.

The final results of F1, F2, and F3 outcomes are the same, but the probability of their occurrence is different, as presented in the following:

$$P(F1) = P1S * P2S * P3F = 0{,}999 * 0{,}9 * 0{,}05 = 0{,}044955 \tag{9.4}$$

$$P(F2) = P1S * P2F = 0{,}999 * 0{,}1 = 0{,}0999 \tag{9.5}$$

$$P(F3) = P1F = 0{,}001 \tag{9.6}$$

The probability of a positive outcome is as follows:

$$P(S) = P1S * P2S * P3S = 0{,}999 * 0{,}9 * 0{,}95 = 0{,}854145 \tag{9.7}$$

The above analysis leads to the conclusion that air conditioning system failure will be resolved with a positive outcome in slightly over 85% of cases:

1. In about 4.5% of the cases, system room temperature monitoring sensor will not register temperature increase above the permitted limit.
2. In almost 10% of the cases, when the sensor registers temperature to increase and the alarm is activated, this will not be noticed by operators who will not react.
3. In only 1% of the cases, it was impossible to activate the redundant unit.

5 FTA Method in the Case of Air Conditioning System Failure

FTA method will deal with the three critical events identified relating to the control mechanisms of the air conditioning system management:

- Temperature sensor does not register temperature increase.
- A member of responsible technical staff does not notice temperature monitoring sensor alarm.
- It is not possible to turn on the redundant unit.

A failure tree will be drafted for each of these cases to register possible reasons for these events. Apart from that, for every event, it is possible to define the probability of its occurrence as well.

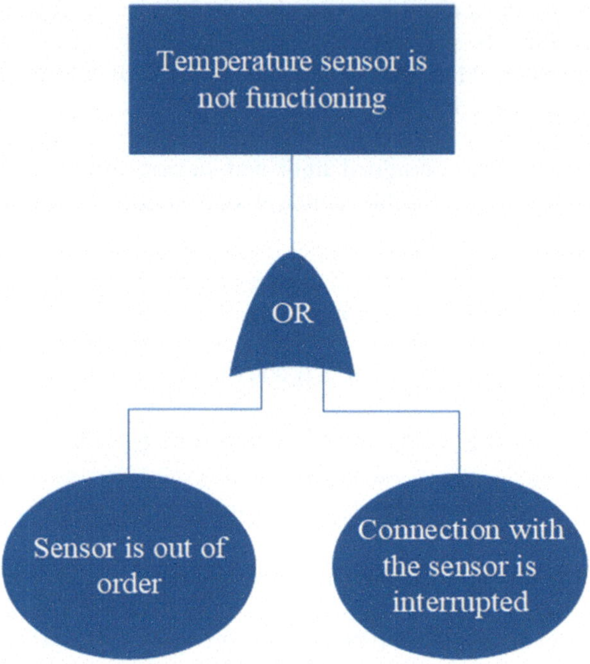

Fig. 9.3 FTA—temperature sensor does not register temperature increase

The temperature sensor in the system room has the task of monitoring the temperature and activating the alarm if the temperature read exceeds the defined permitted limits (Fig. 9.3). The temperature increase is the most common situation registered by the sensor. The reason for which the sensor does not perform its task may be either sensor defect or interruption of the link between the sensor and alarm system.

An alarm system can be composed of a device that gives a sound and/or light signal or something else such as reporting by email or SMS message to the mobile telephone of responsible technical staff members.

It is the task of members of the examined system's responsible technical staff to monitor the alarm system and turn on the redundant unit if necessary. It is possible that the responsible staff does not notice the alarm (Fig. 9.4). The reason can be either an alarm system defect or failure to monitor the alarm.

If needed, i.e., in the case of primary air conditioning unit failure, a responsible technical staff member must turn the redundant unit on once the alarm has been registered. However, there might be cases when it is impossible to turn the redundant unit on (Fig. 9.5). The above diagram shows that among possible reasons there are a defect of the redundant unit, defect of the switch used for turning on the redundant unit in the primary system, or failure to train the appointed responsible technical staff in charge of air conditioning system supervision.

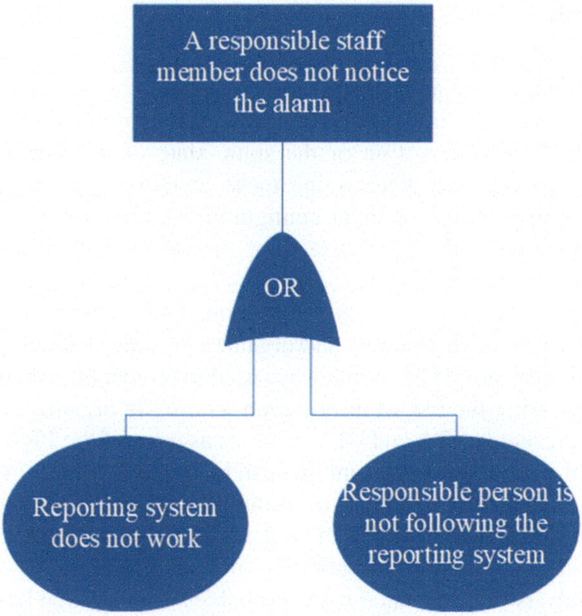

Fig. 9.4 FTA——responsible staff does not notice the alarm

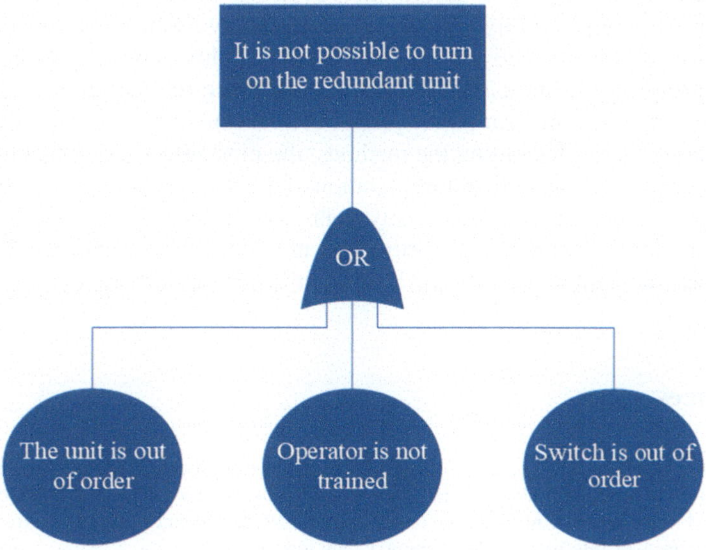

Fig. 9.5 FTA—It is not possible to turn on the redundant unit

6 Conclusion

Risk analysis of undesirable and harmful situations is one of the phases of the safety life cycle. Today, this analysis is carried out from an approach based on the analysis of security controls or observation of the static state of information assets to a scenario-based approach and determining these scenarios' probability. Scenarios reflect system, equipment, component configurations, and critical human–system interactions associated with event detection, diagnosis, mitigation, and system recovery.

The development of information and communication systems leads to a reduction in data security in organizations, and organizations are making great efforts to protect information security. One of the ways used to protect information security is the method of assessing the risk of information security in organizations.

This chapter presented ETA and FTA methods as parts of the PRA methodology in the example of a separate subsystem that can be found in almost every IT center. For the chapter, the initial event (failure of the system room air conditioning system), controls implemented for system protection, possible events related to the implemented controls, and the probabilities of their occurrence were identified. Based on the analysis of the created trees, probable scenarios that have an undesirable impact on the information system's functionality were identified, and additional data were provided for the optimal selection of protection measures.

It follows from the above that the use of PRA methodology and methods used in this chapter are applicable in information systems. Although such methods are demanding, their applicability is possible because today's information systems consist of many subsystems. The level of expected availability of these systems, i.e., the inadmissibility of downtime, has become equivalent to the requirements of high-risk industries such as nuclear power plants. PRA methodology was applied. Given the complexity of implementing the methods, the most effective application of the methodology is in the second or third iteration of the risk assessment, i.e., to be used for a deeper consideration of potentially high risks detected in the initial iteration. This research opens the space for further research in the field of data availability in a virtual environment to increase the level of information security.

References

1. Perakovic D, Perisa M, Cvitic I, Zoric P (2020) Identification of the relevant parameters for modeling the ecosystem elements in industry 4.0. In: Knapcikova L, Balog M, Perakovic D, Perisa M (eds) 4th EAI International Conference on Management of Manufacturing Systems, 1st edn. Springer International Publishing, Cham, pp 111–123
2. Hrvatski sabor Zakon o informacijskoj sigurnosti, NN 79/07
3. Covert Q, Steinhagen D, Francis M, Streff K (2020) Towards a triad for data privacy. In: Proceedings of the 53rd Hawaii International Conference on System Sciences, pp 4379–4387
4. Peraković D, Periša M, Cvitić I, Zorić P (2020) Information and Communication Technologies for the Society 5.0 Environment. In: Zbornik radova trideset osmog simpozijuma o novim

tehnologijama u poštanskom i telekomunikacionom saobraćaju – POSTEL 2020. University of Belgrade, Faculty of Transport and Traffic Engineering, Belgrade, Serbia, pp 203–212
5. Cvitić I, Peraković D, Kuljanić TM (2017) Availability factors in delivery of information and communication resources to traffic system users. In: Mikulski J (ed) Smart solutions in today's transport. Springer International Publishing, pp 28–41
6. Cvitić I, Peraković D, Periša M, Jerneić B (2016) Availability protection of IoT concept based telematics system in transport. In: Mikulski J (ed) Challenge of transport telematics. Springer, Cham, pp 109–121
7. Khosravi-Farmad M, Ghaemi-Bafghi A (2020) Bayesian decision network-based security risk management framework. J Network & Syst Manag 28:1794–1819. https://doi.org/10.1007/s10922-020-09558-5
8. Baskerville R, Spagnoletti P, Kim J (2014) Incident-centered information security: managing a strategic balance between prevention and response. Inf Manag 51:138–151. https://doi.org/10.1016/j.im.2013.11.004
9. Jouini M, Rabai LBA, Ben AA (2014) Classification of security threats in information systems. Procedia Comp Sci 32:489–496. https://doi.org/10.1016/j.procs.2014.05.452
10. International Organization for Standardization (2018) ISO 31000:2018 Risk management — Guidelines. https://www.iso.org/standard/65694.html. Accessed 5 May 2020
11. International Organization for Standardization (2018) ISO/IEC 27005:2018 Information technology — Security techniques — Information security risk management. https://www.iso.org/standard/75281.html. Accessed 5 May 2020
12. Ali N, Hussain M, Hong J-E (2020) Analyzing safety of collaborative cyber-physical systems considering variability. IEEE Access 8:162701–162713. https://doi.org/10.1109/ACCESS.2020.3021460
13. Ketabdari M, Giustozzi F, Crispino M (2018) Sensitivity analysis of influencing factors in probabilistic risk assessment for airports. Safe Sci 107:173–187. https://doi.org/10.1016/j.ssci.2017.07.005
14. Fithri P, Nofriyanti HA, Kurnia I (2020) Risk analysis for occupational safety and health in manufacturing company using FMEA and FTA methods: a case study. IOP Conference Series: Materials Science and Engineering 1003:012073. https://doi.org/10.1088/1757-899X/1003/1/012073
15. Li J, Xu K, Fan B, Geng L (2020) Risk assessment of oxygen lance burning loss using bow-tie analysis based on Fuzzy theory. Math Probl Eng 2020:1–1. https://doi.org/10.1155/2020/2843416
16. Al Saffar IQ, Ezzat AW (2020) Probabilistic risk assessment of typical process plant using event tree and fault tree analysis. J Mech Eng Res Develop 43:465–477
17. Khidzir NZ, Ahmed SAM, T.T. G (2018) Probabilistic risk assessment in A.I enabled SoE attacks. BITARA Int J Civiliz Stud & Human Sci 1:32–39
18. Cvitic I, Perakovic D, Perisa M, Botica M (2020) Definition of the IoT Device Classes Based on Network Traffic Flow Features. In: Knapcikova L, Michal B, Peraković D, Periša M (eds) 4th EAI International Conference on Management of Manufacturing Systems. Springer Nature, pp 1–17
19. Cvitić I, Peraković D, Periša M, Husnjak S (2019) An overview of distributed denial of service traffic detection approaches. PROMET – Traff & Transp 31:453–464. https://doi.org/10.7307/ptt.v31i4.3082
20. Cvitić I, Peraković D, Periša M, Botica M (2019) Novel approach for detection of IoT generated DDoS traffic. Wireless Networks 1–14. https://doi.org/10.1007/s11276-019-02043-1
21. Mosleh A (2014) PRA: a perspective on strenghts, current limitations, and possible improvements. Nucl Eng & Technol 46:1–10. https://doi.org/10.5516/NET.03.2014.700
22. Kwag S, Gupta A (2017) Probabilistic risk assessment framework for structural systems under multiple hazards using Bayesian statistics. Nucl Eng Des 315:20–34. https://doi.org/10.1016/j.nucengdes.2017.02.009
23. United States Nuclear Regulatory Commission Tutorial on Probabilistic Risk Assessment (PRA). https://www.nrc.gov/about-nrc/regulatory/risk-informed/rpp/pra-tutorial.pdf. Accessed 5 May 2020

24. Stamatelatos M (2000) Probabilistic risk assessment: what is it and why is it worth performing it? NASA Office of Safety and Mission Assurance, pp 3–6
25. Ignac-Nowicka J (2018) Application of the FTA and ETA method for gas hazard identification for the performance of safety systems in the industrial department. Manag Syst Prod Eng 26:23–26. https://doi.org/10.2478/mspe-2018-0003
26. Gabriska D (2020) Evaluation of the level of reliability in hazardous technological processes. Appl Sci 11:134. https://doi.org/10.3390/app11010134
27. Waghmode LY, Patil RB (2016) An overview of Fault Tree Analysis (FTA) for reliability analysis. J Eng Res Stud 4:6–8

Chapter 10
Challenges of Improving the Railway Passenger Information System in the Republic of Croatia

Dragan Peraković, Marko Periša, Marjana Petrović, Ivan Cvitić, and Petra Zorić

1 Introduction

The ability to provide information and transport services anytime and anywhere makes the rail network more efficient and easier for end users. Real-time and accurate information is an essential item in informing passengers. The basis for obtaining this information is a stable and comprehensive passenger information system that uses a specific architecture to deliver the service. This system can do real-time transmission of information and data, for example, on the timetable from different sources, their connection into one whole, and providing through various interfaces. The system includes updated information related to any timetable change and timetable changes combined with timetable data and delivered to the information systems. The passenger information system relies on multimedia network technology in which the central computer provides information services to passengers using screens at stations or stops and in vehicles.

Elements of the railway network architecture enable computing capacity and telecommunications to collect, transmit, and process data related to stakeholders' management and information. From the management point of view, the passenger information system is divided into the source of information, central management level, management level in stations and on the train, and control equipment at stations and on the train.

This chapter will present the functionalities of currently available information and communication (IC) solutions and services that enable informing passengers in rail traffic in the European Union (EU). The analysis will include passenger

D. Peraković · M. Periša · M. Petrović · I. Cvitić · P. Zorić (✉)
Faculty of Transport and Traffic Sciences, University of Zagreb, Zagreb, Croatia
e-mail: dragan.perakovic@fpz.unizg.hr; marko.perisa@fpz.unizg.hr; marjana.petrovic@fpz.unizg.hr; ivan.cvitic@fpz.unizg.hr; petra.zoric@fpz.unizg.hr

L. Knapcikova et al. (eds.), *Sustainable Management of Manufacturing Systems in Industry 4.0*, EAI/Springer Innovations in Communication and Computing,
https://doi.org/10.1007/978-3-030-90462-3_10

information systems in the EU countries with the highest realized passenger traffic in the last year, measuring it in billions of passengers transported. Furthermore, these systems will be compared with the Republic of Croatia's (Croatia) passenger information systems. Based on the comparison and analysis of the current state of IC infrastructure, the chapter presents an overview of the possibilities of improving the passenger information system in Croatia.

2 Overview of the Current Passenger Information Systems in the EU

The railway system can be divided into structural area subsystems and functional area subsystems. The structural area subsystem includes infrastructure, energy, monitoring, control and signaling, control orders, and signaling in the vehicle and the vehicle fleet. Furthermore, the functional area subsystem includes traffic management, maintenance, and telematics applications for passenger and telematics services [1]. The passenger information system is a solution for providing relevant real-time information to passengers. It includes a passenger information system, passenger information screens, surveillance, and physical address. It is responsible for the automatic or manually programmed provision of visual and audio data to passengers at stations and stops. Passengers have access to the system through various channels, such as Internet-connected devices, computers, and mobile devices [2].

Mobile application-based passenger information systems are rapidly gaining ground due to the increased penetration of smartphone use. It is most common in passengers who use public transport [3]. Namely, passengers prefer to use data available on mobile devices rather than data displayed on screens at stations or through a public address system [4]. It is necessary to improve the quality of rail passenger transport services so that it reflects customer requirements and optimal use of public resources [5]. Apart from providing passenger information to end user, it is essential to know what real-time information passengers want and what agencies provide—information demand and supply side [6]. Research on existing and future passenger information concepts [7] has shown that innovative and new passenger information technologies represent a great benefit for providers of information services by increasing users' flow in stations. Providing passengers with accurate real-time information can lead to shortening waiting time and decreasing overall travel time due to changes in path choice and increased use of public transport and satisfaction with transit services [8]. The development of IC in the transport sector has resulted in services focusing on customers and their needs by changing passenger transport's previous perception [9]. With the developed technology and accurate information, available passengers can replan their path if any disturbances and disruptions on the network [10]. Modern and efficient railway infrastructure is a prerequisite for the development of rail transport [11]. The application of modern methods and IT tools in railway transport requires additional infrastructure equipment with many technical elements related to data collection, processing, and distribution [12]. Previous research has identified that current railway network modernization projects in the Republic of Croatia mostly include traffic management systems and solutions based on IC infrastructure [13].

By the EU Commission Regulation [14], each station manager must provide train departure data at railway stations to the end user. According to the Regulation, the telematics application subsystem consists of two elements: a passenger transport application and a freight application. Passenger transport applications further include systems for providing information to passengers before and during the journey, reservation and payment systems, baggage management, and the management of train connections and other modes of transport. The Regulation is based on the Technical Specifications for Interoperability (TSI) in Directive 2001/16/EC of the European Parliament and of the Council of March 19, 2001 on the trans-European interoperability conventional rail system, which details the provision of passenger information and the implementation of the passenger information system. The TSI aims to establish procedures and interfaces among all stakeholders to provide relevant information and issue tickets to passengers via currently available technologies [15]. The railway carrier is obliged to provide information from its timetable to other railway undertakings and third parties. All relevant information and timetable data must be accurate and up-to-date, and available to other railway undertakings and third parties within the next 12 months after they expire and become invalid. According to the EU Regulation, the basic guidelines for developing TSIs and their harmonization with European standards are defined. In 2020, the Croatian Parliament passed a law that is important for harmonizing information and the railway system [16].

A 2020 Eurostat survey on the impact of rail passenger transport at the EU level shows that the largest number of passengers in 2018 was carried in Germany (2,880,558,000), the United Kingdom (1,783,232,000), and France (1,246,804,000) [17]. For this reason, passenger information systems have been analyzed in the area of railway administrations of these countries. The railway system in these countries is much more complicated, so modern IC technology is integrated into all subsystems in order to maintain a functional and efficient system. In many EU countries, such an approach in the processes of modernization and introduction of modern communication solutions and services in the railway system has enabled the efficient, reliable, and safe operation of the system.

The current passenger information systems on the rail network in Germany, the United Kingdom, and France, which are developed according to the TSI guidelines and the EU Commission's Regulation, can be divided into passenger information systems before and during the trip. Passenger information systems before the trip include information via the website and mobile applications. During the trip, passenger information systems refer to audio and visual information via screens in the stations themselves or at the stops.

The German railway network infrastructure (DB Netz AG) is managed by the German Railways (Deutsche Bahn AG). It manages 33,300 [km] of the railway network, 5400 stations, and stops. The passenger information system used at stops and stations is fully developed in the environment of Cloud Computing technology [18]. Siemens solutions are one of the most represented on the railway network in Germany. Information provided to users at stations and stops is available in video and audio format. In addition to visual systems, the passenger information system also uses voice information. For this purpose, a fully automatic sound system "Voice over IP" is used, which projects voice announcements. Route number, destination station, and connecting transit are announced on tact, while delays, cancellations, closed doors, or route changes are announced as needed. General announcements regarding the ban on smoking and unattended luggage are also published regularly. An essential feature of this passenger information system is the ability to manually operate the entire system to provide important information to passengers in an emergency.

The UK rail network is managed by Network Rail Infrastructure Ltd. and has a network of 32,186 [km] of the track. The system for informing passengers at stations and stops includes the CIS—user information subsystem and OIS—operational information subsystem. CIS LED/LCD screens are located at the entrances to stations and/or platforms and provide departure/arrival information. The CIS subsystem also allows communication with arriving trains. OIS screens provide additional contextual or general information (for example, informing passengers about planned works) and possibly additional information during traffic jams. The solution represented on the UK railway network is the Infotec information system and is based on modern communication technologies, which enables the processing of all relevant data and real-time information to passengers. The technologies used by the system are Cloud Computing (Amazon AWS), IoT, 3G, 4G, and GPRS [19].

The railway network in France is managed by the SNCF group (Réseau) and consists of a network of 30,000 [km] of the track, of which 2600 [km] refers to high-speed rail [20]. Televic Rail's solution is used in vehicles, while Teleste, Alstom, and Thales group solutions are used at stops and stations. The Thales group's solution consists of signaling, video surveillance in vehicles and at stations, and a passenger information system. It is based on Cloud Computing technology, Big Data, and artificial intelligence. The services being developed in this environment provide passengers with information via video, audio, the multimedia system at stations and stops, and mobile devices. Elements of the IC system architecture are sensors, GPS devices, network components (routers, LAN, and WiFi), Cloud Computing, and more.

3 Comparison of the Passenger Information System in Croatia with the EU

The passenger information systems applied in the countries involved in the analysis of work are the best example of the implementation of passenger information practices in railway transport. To position the passenger information system's possibilities and the level of passenger information availability in Croatia to the EU, a comparison of this system's features with the previously analyzed systems will be presented.

On the rail network of the infrastructure manager in Croatia, HŽ Infrastructure (HŽI), out of a total of 545 stations and stops in Croatia, only 256 are equipped to provide visual information for passengers. Of the 124 stations equipped with the listed equipment required to provide information, six stations also provide dynamic screens for displaying information to passengers. There are 60% of stations and stops with the possibility of providing visual information to passengers, with only about 1.4% of them having dynamic screens [21].

The service of informing passengers at stations and stops is provided via loudspeakers or visually via fixed bulletin boards displaying the all-day timetable, i.e., screens with information on the time and place of train arrival and departure, train delays, possible change of transport route, and other necessary information related to rail traffic. Data on railway lines exist in digital form on the passenger transport operator's website and the mobile application.

Most stations and stops have a passenger information system consisting of several subsystems [22]:

- Classic fixed timetable
- Boards (led screens)
- Clock subsystems
- Loudspeaker subsystems
- Ups (uninterruptible power supply)
- Communication equipment that allows it

Analog clock screens located at stations and stops are composed of SMD (Surface-Mount Device) LED light modules designed to display various alphanumeric messages in multiple lines at rest. Each side of this screen consists of a display part (LED screen), a processor assembly, a power supply module, a communication interface, a light sensor, and lighting and switching assemblies. The information is created in a control center that sends it to the processor assembly, where it is processed and displayed on an LED screen. The communication interface is optically separate and connects the information screen and the control computer. The lighting and switching assemblies' sensors illuminate the mechanical clock and the inscription on display.

The clock subsystem consists of a master clock developed in modular technology, a two-wire transmission system (enables independent installation of auxiliary clocks without maintenance and a remotely synchronized computer system) MOBAWNT software program. The loudspeaker subsystem consists of a digital output module, a power amplifier, a universal interface module, a digital telephone exchange/digital key module, and a converter. The UPS subsystem is designed to prevent interruptions and impacts on computers and valuable electronic equipment. Communication equipment in stations are numerous and can be divided into the following elements: media converters, relay modules, power relay modules, input modules, smart web input/output Ethernet module, interface converter, various connectors, data cable shielding, and industrial server serial devices over a TCP/IP-based Ethernet network.

Table 10.1 Comparison of passenger information systems during the trip

	DB	NRN	SNCF	HŽI
Visual informing	+	+	+	+
Voice informing	+	+	+	+
Dynamic screens	+	+	+	+
Touch screens	+	+	+	–
Type of screens	TFT-LCD	LED	TFT	LED
Pointer direction	+	+	+	–
Diagnostic feature built into the screen	+	–	–	–
Time and date display on the screen	+	+	+	+
Display of train arrival/departure on the screen	+	+	+	+
Promotional content on screens	+	+	+	–
Adapted for disabled people	–	–	+	–

Table 10.2 Comparison of passenger information systems before the trip websites

	DB	National rail	SNCF	HŽPP
Availability on more languages	+	–	+	+
Adaptation for the visually impaired persons	+	–	+	–
Enabled direct purchase of tickets through the website	+	–	–	+
Multiple ticket payment methods enabled	+	+	+	+
Obligation to register and register when buying tickets	+	+	+	–
Seat reservation enabled	+	+	–	–
Travel planner	+	+	+	+
Benefit overview	+	+	+	+
Available information related to bicycle and pet transport services	–	+	+	+
Timetable display	+	+	+	+
Information about timetable changes	+	+	+	+
Passenger rights information	+	+	+	+
Canceling a ride or changing the ticket directly on the page available	+	–	–	–

The technology used and how information is displayed to passengers at stations and stops in the analyzed countries differ. Table 10.1 shows the main differences between passenger information systems during the trip.

According to a comparison of the passenger information system features during the trip, the systems used in Germany and France provide the most opportunities for informing passengers. On the other hand, informing passengers in Croatia is limited only to the necessary information, such as displaying time and date and departure time by visual or audio means. It should be noted that the above information is not available to passengers in all stations and that LED screens are in a limited number of large stations.

Table 10.2 shows the differences between the services offered through websites in the analyzed countries. Comparing the possibilities of informing passengers via

Table 10.3 Comparison of mobile applications of passenger information systems before trip

	DB navigator	National rail	SNCF app	HŽPP planer
Availability in multiple languages	+	–	+	+
Availability for devices on Android, IoS, and Windows phone platforms	+	+	–	–
One application that contains all the services	+	+	–	–
Enabled ticket purchase directly through the application	+	–	–	–
More ways to pay for tickets	+	+	+	–
Possibility to book a ticket	+	–	–	–
Possibility to reserve a seat	+	–	–	–
Travel planner	+	+	+	+
Offline map display and traffic information	–	–	+	+
Driving time schedule	+	+	+	+
Notices related to timetable changes	+	+	+	+
Additional offers	+	+	+	+
Adaptation for people with visual and hearing impairments	–	–	+	–

websites among the considered infrastructure managers, it is evident that DB provides the broadest range of services to passengers. In Croatia, the possibility of booking a seat, canceling, or exchanging tickets and registration when buying tickets is not available and is directly related to the passenger carrier. These shortcomings are not justified, given that there is still only one passenger transport operator in Croatia. In other countries, there are many more due to the liberalization of the passenger transport market.

The mobile application for informing passengers HŽPP Planer, among other things, offers the possibility of calculating the price of a ticket (with discounts, if the passenger is entitled to them), tracking the GPS position of trains, travel details, and the similar. Table 10.3 shows the essential characteristics of the analyzed applications.

According to the parameters listed in Table 10.3, it can be concluded that the mobile application for passenger information DB Navigator satisfies the broadest range of parameters considered.

4 Guidelines for the Improvement of the Passenger Information System in the Railway Transport of the Republic of Croatia

In 2019, a survey was conducted regarding the quality of railway passenger service in Croatia. Among other service elements, information availability was the focus of mentioned research. The survey results show that "information availability regarding service cancellation" was insufficient by 50% of users. "Available information regarding timetable changes" is rated insufficient by more than 35% of the users. Other elements regarding information availability were rated better [23].

According to the EU Regulation, the basic guidelines for developing TSIs and their harmonization with European standards are defined. The modernization of the passenger information system's objectives is based on the laws mentioned above and EU Regulations. The goals of modernization are

- Improving and maintaining the quality of railway transport services
- Ensuring compatibility with existing subsystems
- Compatibility of high-speed railway systems and conventional railway systems depending on the category of lines
- A railway system accessible to all regardless of the degree of disability
- Improving economic efficiency
- Mobility of system users
- Increasing the competitiveness of the Croatian railway measure
- User information

An IC architecture is needed to deliver accurate, secure, relevant, and real-time information to end users. Figure 10.1 shows a proposal for such an architecture. It can be seen that the passenger information system is part of the structure of the operational management and traffic control center. In this proposal, the data center is equipped with IoT sensor systems to monitor and control all parts of the system whose possibility of application in the railway system is the subject of research in [11]. It is necessary to pay attention to the possibility of cyber threats such as DDoS (Distributed Denial of Service) attacks on the system [24–26]. In such a structure, the current development direction is taken into account, i.e., integrated passenger transport projects [27].

In such a complex system, three subsystems listed and described in Table 10.4 participate in creating and delivering information to end users on the railway network. Currently, there is only one passenger transport service provider (HŽPP) in Croatia. However, this chapter's proposal enables any passenger carrier's connection for a specific category of users. Each of these subsystems can be modernized. For this chapter, the authors focused on the modernization of the infrastructure subsystem (HŽI).

$$S(R) = f\{S_{\mathrm{I}} \,?\, S_{\mathrm{PTR}} \,?\, S_{\mathrm{IT}}\}, \tag{10.1}$$

where

S_{I}—infrastructure subsystem
S_{PTR}—passenger transport by rail subsystem
S_{IT}—integrated transport subsystem

A communication network and a security concept must be used to exchange data to enable any network transmission. The information exchange architecture favors

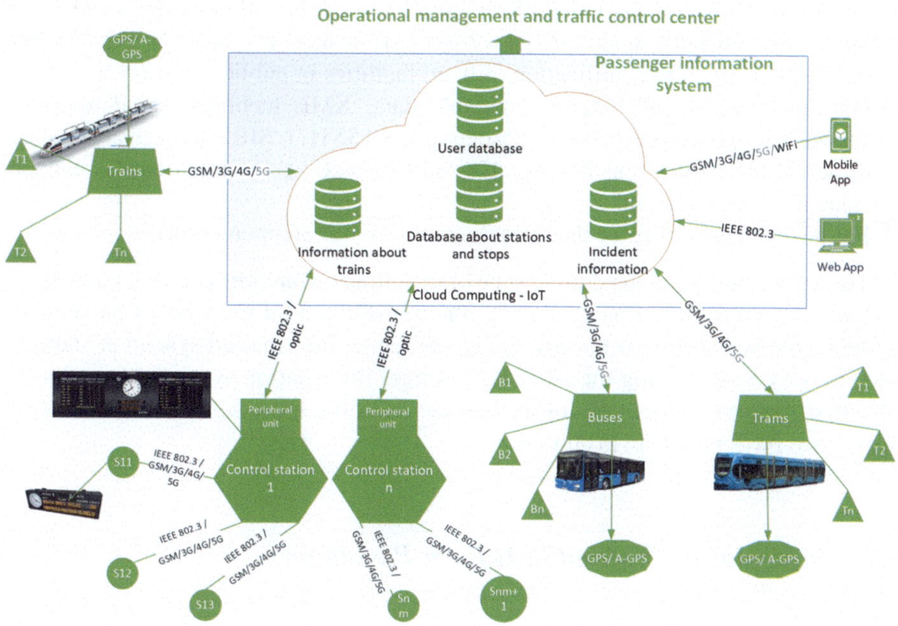

Fig. 10.1 Proposal of IC architecture for information delivery [28]

Table 10.4 Subsystems that are part of a system for delivering information to passengers on the railway network

Subsystem	Subsystem parts	Acronym
S_I	HŽI	$S_{HŽI}$
S_{PTR}	HŽPP	$S_{HŽPP}$
	Passenger transport operator 1	S_{PTO1}
	Passenger transport operator n	S_{PTOn}
S_{IT}	ZET	S_{ZET}
	Operator 1	S_{O1}
	Operator n	S_{On}

peer-to-peer interaction between all participants while guaranteeing the interoperable railway community's overall integrity and consistency by providing a full range of centralized services. The railway system's interoperability is based on a typical "information exchange architecture" known and adopted by all participants, thus encouraging the reduction of barriers for new participants, especially service users. The central data container is first accessed to access the meta information, for example, to verify the security data. Peer-to-peer communication then takes place between the participants involved. In order to manage other modes of transport, the following standards apply to the provision of information and the exchange of information with other modes of transport:

- For the exchange of timetable information between the railway carrier and other modes of transport: standard EN 12896 ("Transmodel") and EN TC 278 WI 00278207 ("IFOPT identification of fixed facilities in public transport")
- For the exchange of specific timetable data, XML technical standards, and Transmodel-based protocols, in particular EN 15531, ("SIRI"), for the exchange of actual timetables and EN TC 278 WI 00278207 ("IFOPT") data exchange standards "Stop/station"
- For the exchange of price data (this standard is still an open point)

The authors analyzed the current state of the infrastructure for receiving equipment and all relevant information needed for the transfer and information of passengers to define the possibility of improving the passenger information system at stations and stops by modernizing the current passenger information system. Also, phases of introducing the system for official places with the most significant number of dispatched passengers are proposed.

4.1 Information Needed to Inform Passengers

The information needed to inform passengers needs to be viewed from several different interactions between the relevant stakeholders:

- Between mobile units (vehicles) and central system units (servers)
- Between information screens at stations/stops and central system units
- Between the central units of the system and the vehicles and information screens at the stations/stops
- Between central and peripheral units

Data related to current train status, traffic regime on the routes of individual lines, driving time, distance traveled, the vehicle's status, drivers, lines, timetable, passengers, and driving units are exchanged between the vehicle and the server. Information screens at stops/terminals and servers exchange data such as the time until the arrival of an individual vehicle at the stop, the status of the information screen at the stop, the possibility of combining different modes of transport, traffic conditions.

Central system units, vehicles, and information screens at stops exchange data of different trip planners, different text and/or voice notifications and traffic information intended for drivers, traffic staff or passengers, and georeferenced information of all system users (driving/transport units, information screens at stops, other vehicles, etc.) according to the transport system, individual line, individual vehicle unit, and individual information display. They also exchange necessary information when changing routes (length, number, and arrangement of stopping points), driving time (turnover/half-turn/stopping at stops, standing at terminals, time of day, types of days), sequence, frequency, capacity, type and sort of vehicle, traffic area/zones, and information on changing routes intended for passengers, in vehicles and at stops, and on various web (mobile) applications.

Central and peripheral units interact with each other regarding the exchange of timetable data and various timetable elements (line, stops, planned/actual departures-arrivals per stop, individual journeys, driving duration, driving/transport units), staff data/workers, vehicles, line routes, stops/terminals, timetables, and all other information of interest.

4.2 The Existing State of Infrastructure for Receiving Equipment

The chapter presents an in-depth analysis of available data by HŽI related to certain official places' IC infrastructure equipment (Table 10.5). At some official places in Croatia, there are complete passenger information systems manufactured by EXOR 2002, 2009, 2011, 2014, ETC digitalpro, and Elektrokem. The system elements in the form of clocks and loudspeakers are placed in places where complete systems are not located. The existence of cable ducts is also present at specific locations and access points important for connecting an official location, such as a DuoTrack cable. In addition to the above, there is an STA or STKA cable (signal telecommunication cable with aluminum sheath and coaxial pairs) and a TD-TF cable (copper conductors) on the railway network. The table below is essential for understanding the current state of the infrastructure. It can be seen that only larger official places,

Table 10.5 Analysis of the current state of the IC infrastructure for several official places

Official place name	Status	Main station	Existing PIS	Cable canalization	Connection with official places (rail cable)	Electro energetic cable
M102 Zagreb GK-Dugo Selo						
Zagreb Glavni Kolodvor	Main station		YES/ EXOR/2002	YES	YES/STKA	YES
Maksimir	Station	Zagreb Borongaj	NO	NO	YES/TD	YES
Zagreb Borongaj	Main station		NO	YES	YES/STKA	YES
Trnava	Station	Zagreb Borongaj	NO	YES	YES/TD	YES
Čulinec	Station	Zagreb Borongaj	NO	NO	YES/STKA	YES
Sesvete	Main station		NO	NO	YES/STKA	YES
Sesvetski Kraljevec	Station	Sesvete	NO	NO	YES/STKA	YES
Dugo Selo	Main station		NO	YES	YES/STKA	YES

such as Zagreb Glavni Kolodvor, have an existing complete passenger information system, while smaller official places do not have any passenger information systems.

Optical backbones are the basis for future modernization in providing real-time passenger information services and other upgrades of the IC system. The modernization of the IC system began in 2015, and to this day, the old technological system has mainly been replaced by a new one. Telecommunication devices have been partially modernized on the Croatian corridors (former corridors X, Vb, Vc, and branch Vb1). This refers to installing an optical cable, the SDH backbone, the replacement of the ŽAT exchange with a new digital one, and the construction of a new IP/GBE data transmission network.

4.3 Stages of Introduction of the Passenger Information System

For this chapter, identifying phases of development and improvement of the passenger information system were performed. The authors analyzed the first ten official places that dispatched the largest number of passengers in 2019 in the Republic of Croatia. The met technical minimums in terms of construction and signal safety were also taken into account. The phases of introducing the passenger information system are based on the plans for the development and modernization of the railway infrastructure and the projects in progress. They are proposed according to the state of the IC infrastructure on the railway network in Croatia and according to supply and demand analysis. Table 10.6 shows whether the analyzed official places with their current information and communication infrastructure meet the technical characteristics to introduce the passenger information system. The condition for satisfaction is set if there is a cable sewer, access point, a railway conductor, and an electricity connection at the official place.

Phases 1 and 2 represent the implementation of existing projects under construction or planned to be implemented shortly. For certain official places, those that do not meet the minimum technical and technological requirements, the necessary measures for introducing a passenger information system are listed. Although this is a railway modernization plan, it is possible to include measures for equipping the passenger information system in the plans due to the adequate IC infrastructure.

After fulfilling all the technical preconditions, it is possible to start the phasing of introducing the information system at the stations and stops. Implementing the passenger information system should be carried out in phases, which enables certain parts of the project to be more efficiently implemented and the possibility of making the right decisions on further strategic directions for the modernization of the entire Croatia railway system.

Table 10.6 Review of the analysis of necessary measures for official stations that do not meet the requirements for the introduction of the passenger informing system

No.	Official place	Total dispatch	Telecommunication network meets the requirements	Implementation phase	Necessary measures
1	Zagreb GK	4,207,616	YES	1 and 2	
2	Dugo Selo	819,723	YES	1 and 2	
3	Sesvete	700,098	YES	1	
4	Zabok	603,106	YES	1	
5	Vrapče	480,937	BAD	3	Modernization of the telecommunications network
6	Zagreb ZK	453,791	YES	No data	
7	Varaždin	437,458	YES	1	
8	Gajnice	396,726	BAD	3	Connection of the official place with the rail cable
9	Osijek	393,803	YES	3	
10	Sl. Brod	386,148	YES	3	

4.4 *Opportunities to Improve Passenger Information System in Croatia*

Given that passengers today need to obtain specific information related to travel and the possibility of choosing such information, it is necessary to equip stations and stops with solutions that have touch screens to achieve interactivity with passengers. Promotional materials can also be placed on such screens while passengers are not using them, and they could be adapted for people with visual and hearing impairments. Improving the passenger information service at stations is possible by incorporating diagnostic features into the displays. In this way, the passenger operator could react promptly to any interference, which would raise the quality of information delivery.

The most significant current disadvantage is visible in providing information to people with visual and hearing impairments, as the screens at the stations are not adapted to these groups of users. The mentioned shortcoming is also visible when informing passengers before the trip, i.e., informing via the website and mobile applications. The information that the railway undertaking must provide to users with visual or hearing impairments must comply with the WCAG (Web Content Accessibility Guidelines) 2.1 guidelines. Also, the assistance reservation service provided by the railway undertaking should be based on modern communication solutions.

For the time being, the TAP-TSI is used to exchange data in rail transport. It is essential from an integrated passenger transport aspect (a combination of rail and other forms of public passenger transport), and it is essential to avoid duplicating travel data that may provide the passenger with inaccurate information. As a result, it could endanger the safety of passengers, especially if they are people with disabilities for whom the independent movement of the transport network can, in many cases, cause problems [29–31]. Therefore, it is essential to create a national access point to provide accurate and real-time information to passengers [32]. This point relies on a digital interface that provides relevant data and metadata from all transport service providers. In this sense, the provision of information is based on the CEN NeTEx CEN/TS 16614 data exchange standard based on the basic conceptual data exchange model Transmodel EN 12896: Version 2006 (public transport where railway transport is partly included). Combining existing public and private access points into one point would further increase the availability of accurate and real-time information to passengers and increase the level of customer satisfaction with the requested service.

Current ways of informing passengers in rail transport in Croatia face several challenges. As previously seen, it is necessary to modernize the way passengers are informed at stations and stops. By implementing the new information system, the challenges and problems listed in the previous section can be entirely or almost eliminated. Thus, it is needed to monitor modern IC solutions and services and within the defined TSI recommendations to implement optimal solutions on the railway network.

5 Conclusion

Advances in digital technology and the way data is used drastically, changing aspects of society, including industrial production, the private lives of individuals, and traffic in general. Thanks to it, travelers can more easily get all the necessary information related to travel planning. Travel planning is essential for passengers, with IC solutions and services playing a pivotal role. Thanks to the development of these solutions and services, passengers are provided with developed information systems.

Germany, the United Kingdom, and France are at the top of the EU regarding the number of passengers transported by rail per year. Their passenger information systems do not differ too much in the offer of services. On the other hand, currently available passenger information systems in Croatia are still not enviable. The modernization of IC equipment at stations and stops is indispensable regarding technological solutions developed daily. With the modernization of IC equipment, more satisfied passengers are expected to use the railway more often as a form of transport to go to work or school and as the primary form of transportation to more distant destinations.

According to the available data on the IC infrastructure's state on the railway network in Croatia, the passenger information system can be implemented according to the introduction phases and system upgrade measures. For construction requirements, it is necessary to satisfy all the prescribed articles that regulate arranging the station building and the station.

This chapter is the basis for further research and development of passenger information systems on the railway network in Croatia and its integration with other railway management subsystems such as ticket management adaptive for individual user groups. It also opens the possibility of comparing data with other passenger information systems in Eastern and Central Europe's neighboring countries. In addition to the above, future research will be based on the improvement of detected difficulties in the existing system to improve the information service to end users at the national level.

References

1. European Commission (2008) Directive 2008/57/EC of the European Parliament and of the Council of 17 June 2008 on the interoperability of the rail system within the community
2. Allied Market Research (2018) Passenger information system market
3. Lyons G, Jain J, Weir I (2016) Changing times – a decade of empirical insight into the experience of rail passengers in Great Britain. J Transp Geogr 57:94–104. https://doi.org/10.1016/j.jtrangeo.2016.10.003
4. Transport Focus (2014) Rail passengers' priorities for improvements
5. Dolinayova A, Cerna L (2020) The possibilities of increasing the economic efficiency of regional rail passenger transport—a case study in Slovakia. In: Marinov M, Piip J (eds) Sustainable rail transport. Springer, Cham, pp 97–127
6. Harmony XJ, Gayah VV (2017) Evaluation of real-time transit information systems: an information demand and supply approach. Int J Transport Sci Technol 6:86–98. https://doi.org/10.1016/j.ijtst.2017.05.003
7. Peña Miñano S, Kirkwood L, Court S, Farnsworth M, Orlovs I, Shehab E, Tinworth N (2017) A review of digital wayfinding technologies in the transportation industry. In: Advances in transdisciplinary engineering. IOS Press BV, Amsterdam, pp 207–212
8. Brakewood C, Watkins K (2019) A literature review of the passenger benefits of real-time transit information. Transp Rev 39:327–356. https://doi.org/10.1080/01441647.2018.1472147
9. Camacho TD, Foth M, Rakotonirainy A (2013) Pervasive technology and public transport: opportunities beyond telematics. IEEE Pervasive Comput 12:18–25. https://doi.org/10.1109/MPRV.2012.61
10. Zhu Y, Goverde RMP (2019) Dynamic passenger assignment for major railway disruptions considering information interventions. Netw Spat Econ 19:1249–1279. https://doi.org/10.1007/s11067-019-09467-4
11. Shao X, Liu JJ, Gao ZM, Wang P (2012) A study on intelligent onboard rail passenger service systems. In: Ni YQ, Ye X (eds) Proceedings of the 1st international workshop on high-speed and intercity railways. Springer, Berlin/Heidelberg, pp 69–82
12. Kornaszewski M, Pniewski R (2019) The impact of new telematics solutions on the safety of railway traffic on the example of modern simulators railway traffic control devices. In: Mikulski J (ed) Development of transport by telematics. Springer, Cham, pp 32–43
13. Matulić I, Musa M, Peraković D (2014) Information and communication infrastructure for the organisation of railway passenger transport. In: Čokorilo O (ed) Proceedings of the sec-

ond international conference on traffic and transport engineering ICTTE. City Net Scientific Research Center Ltd, Belgrade, pp 410–419
14. European Commission (2011) COMMISSION REGULATION (EU) No 454/2011 of 5 May 2011 on the technical specification for interoperability relating to the subsystem 'telematics applications for passenger services' of the trans-European rail system (Text with EEA relevance)
15. Croatian Parliament (2015) Zakon o sigurnosti i interoperabilnosti željezničkog sustava
16. Croatian Parliament (2020) Zakon o sigurnosti i interoperabilnosti željezničkog sustava
17. Eurostat (2020) Rail transport of passengers. https://appsso.eurostat.ec.europa.eu/nui/show.do?dataset=rail_pa_typepas&lang=en. Accessed 21 Mar 2020
18. Primcore Railway Company uses digital signage for instant updates. https://pimcore.com/en/customers/deutsche-bahn-ag_c675. Accessed 18 Jan 2021
19. Infotec – PIS. http://www.infotec.co.uk/index.php?page=infotec-introduction. Accessed 16 Jan 2021
20. SNCF Réseau. https://www.sncf.com/en/group/profile-and-key-figures/about-us/who-we-are. Accessed 16 Jan 2021
21. Republika Hrvatska (2017) Nacionalni provedbeni plan za Uredbu Komisije (EU) 1300/2014 od 18. studenoga 2014. o tehničkoj specifikaciji za interoperabilnost u vezi s pristupačnošću željezničkog sustava Unije osobama s invaliditetom i osobama s ograničenom pokretljivošću
22. Zorić P, Mikulčić M, Musa M, Kuljanić TM (2021) Analysis of available information and communication solutions and Services for railway passenger information in the EU. In: Knapčíková L, Peraković D, Behúnová A, Perisa M (eds) 5th EAI international conference on management of manufacturing systems. Springer International Publishing, Cham
23. Solina K, Škrtić I, Šimunec I (2020) Analiza savjetovanja s predstavnicima korisnika usluga željezničkog prijevoza tereta i putnika. Željeznice 21:1
24. Cvitić I, Peraković D, Periša M, Botica M (2019) Novel approach for detection of IoT generated DDoS traffic. Wirel Netw:1–14. https://doi.org/10.1007/s11276-019-02043-1
25. Cvitić I, Peraković D, Periša M, Husnjak S (2019) An overview of distributed denial of service traffic detection approaches. PROMET Traffic Transportation 31:453–464. https://doi.org/10.7307/ptt.v31i4.3082
26. Pawlik M (2019) Railway safety and security versus growing cybercrime challenges. In: Mikulski J (ed) Development of transport by telematics. Springer, Cham, pp 57–68
27. Integrirani promet zagrebačkog područja. http://www.ipzp.hr/. Accessed 5 Feb 2020
28. Peraković D, Periša M, Petrović M, Cvitić I, Zorić P (2020) Studija implementacije sustava informiranja putnika na željezničkoj mreži. HŽ Infrastruktura, Zagreb
29. Peraković D, Periša M, Cvitić I, Brletić L (2018) Innovative services for informing visually impaired persons in indoor environments. EAI Endorsed Transactions on Internet of Things 4:156720. https://doi.org/10.4108/eai.5-3-2019.156720
30. Periša M, Cvitić I, Peraković D, Husnjak S (2019) Beacon technology for real-time informing the traffic network users about the environment. Transport 34:373–382. https://doi.org/10.3846/transport.2019.10402
31. Periša M, Kuljanić TM, Cvitić I, Kolarovszki P (2019) Conceptual model for informing user with innovative smart wearable device in industry 4.0. Wirel Netw. https://doi.org/10.1007/s11276-019-02057-9
32. Mlinarić TJM, Ševrović M, Kasum J, Štimac I, Gregurić M, Mandžuka S, Vidović K, Nikšić M, Peraković D, Grgurević I, Škorput P, Periša M, Vojvodić S, Jakovljević M, Beganović D, Lale O, Ljubaj I, Mikulčić M, Civadelić Z, Sikavičev K, Kljajić D, Dujak M (2019) Studija za pružanje multimodalnih informacija u prometu. HŽ Infrastruktura, Zagreb

Chapter 11
Spammer Detection Approaches in Online Social Network (OSNs): A Survey

Somya Ranjan Sahoo, Brij B. Gupta, Dragan Peraković, Francisco José García Peñalvo, and Ivan Cvitić

1 Introduction

In day-to-day life, taking the benefit of Web 2.0 people uses e-commerce and opinion-sharing web applications for information sharing and communication. These websites allow the users to share their emotions, attitudes, personal experiences, feeling regarding products and services, and issues related to politics and economics. In recent years, the review of some specific products or websites increases dramatically. The reverse purchase decision depends on posted opinions by various social network users. Spam refers to unsolicited messages that spread over the network through emails, and direct messages sent by instant messenger, social networks, and various web-based searches depicted in Fig. 11.1. By taking the advantage of these services, spammer spreads malicious contents over the network in the form of malware and phishing [1–3]. Initially, spam spreads and is targeted to limited communications like email and instant messaging. But, it effectively invaded all media across WWW. Spam email called junk mail spreads through unwanted messages or bulk messages with commercial content. Similarly, instant message services like Yahoo Messenger and Skype were used by the adversary to

S. R. Sahoo
Vellore Institute of Technology, Amaravati, Andhra Pradesh, India

B. B. Gupta (✉)
National Institute of Technology, Kurukshetra, India

D. Peraković · I. Cvitić
University of Zagreb, Zagreb, Croatia
e-mail: dperakovic@fpz.unizg.hr

F. J. G. Peñalvo
University of Salamanca, Salamanca, Spain
e-mail: garcia@usal.es

L. Knapcikova et al. (eds.), *Sustainable Management of Manufacturing Systems in Industry 4.0*, EAI/Springer Innovations in Communication and Computing,
https://doi.org/10.1007/978-3-030-90462-3_11

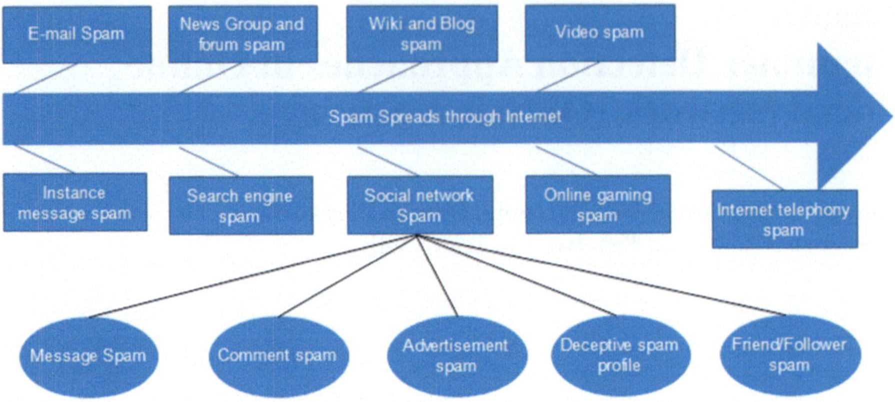

Fig. 11.1 Types of spam spread over the Internet

spread malicious information directly to annoy users. Using short messaging services, mobile-based spammers spread malicious information and infect mobile devices. For the promotion of a particular web page or web application, the spammer manipulates the ranking of search engines and some relevant algorithm. In another way, using short URL-based services, spammer spreads malicious information inside certain blogs and comments over internet services. Social media website like YouTube is the appropriate way for spammers to spread malicious videos with some pornographic and dating websites. The user comments related to those videos are spread over the wired and wireless networks and attract many users to visit [4–6]. Sometimes, these comments are auto-generated through a bot and invite people to surf. Even if the prominent way of communication among different users like social network using Cloud and other ways are also affected by spammers to gain users' credentials [7–19].

Recently, spam inevitably in almost all forms of communication and damage user's content including the performance of the network. It represents one of the biggest security and system performance problems together with DDoS (Distributed Denial of Service) attack [20–22]. Various solutions have been measured to detect spammer content and to improve the performance of the network. These solutions are well known as anti-spamming techniques or spam combating techniques. While a lot of work has been done in the area of malicious content detection based on spammer analysis especially web-based, email spam, spam in social networks and social media, is not even analyzed. This is because of the uncontrollable structure of the social network and flooded content of information. Due to the conductive breeding of social network and the large set of user's activities, it leads to hues damage to mankind. According to various surveys by different companies and brand protectors, spam content increases rapidly day by day. According to other surveys, the growths of spam are rampant. With time, the number of users of social media increases according to the user requirement and communication. As spammer content hampers the performance of the user content and communication medium is associated with the financial loss that is causing erosion in the user behavior. All these factors motivate us to work on social media spam and its detection mechanism.

Due to the unlike behavior of social media platforms based on characteristics, detection of the spammer is challenging and multifaceted. A number of approaches have been developed by researchers and academicians to fight against social spammers including the protection mechanism inbuilt with social network websites. A brief overview of different social spam detection techniques is depicted in Fig. 11.5. However, due to the fastest growing social network platforms, the behavior of users changed rapidly in the last few years. Anti-spam schemes need a major upheaval to extenuate them. In this chapter, we survey various mitigation and detection frameworks that have been proposed in the last few years to fight against spam in OSNs. The rest of the chapter is organized as follows. In Sect. 2, we describe the overview of different spam related to OSNs. In Sect. 3, we elaborate on the types of spammer and various detection mechanisms. In the next section, i.e., in Sect. 4, we describe a literature survey on various spam detection approaches and features of the Microblogging platform by which spammers propagate. In Sect. 5, we elaborate on various comparative analyses with existing approaches. Finally, in Sects. 6, we discuss some open issues and challenges related to spammer detection which concludes the chapter.

2 Online Social Network Spamming

Social network spammers spread in various ways such as posting malicious URLs, short URLs, fake advertisements for publicity, malware spreading, botnet attack through users and systems [23], following unknown users randomly, and some other ways to flooded network [24]. Another method of spreading spammer is the generation of fake reviews of various products and services using machine learning approaches [25]. The growth of global spam increases rapidly over the year and affects every social platform. Specifically, on Twitter one spam is found in every 20 tweets and posts. Most of the spammer content can spread automatically using the system through a bot [26]. Due to the lack of physical contact between the individuals, growth rate of spam increases. Due to these activities, identification of the user is under the black box. Evidently, utilizing the social network data without filtering the malicious activity for analysis is a wrong pattern for social network users. Numerous approaches have been developed by researchers and corporate analysts depicted in Sect. 3. However, spammers develop quickly to evade detection systems.

2.1 Types of Spammer and Spreading Techniques

Spammer spreads over the social network based on the features, properties, and characteristics of various accounts associated with service providers. Various categorizations of spammers spread over social networks are depicted in Fig. 11.2.

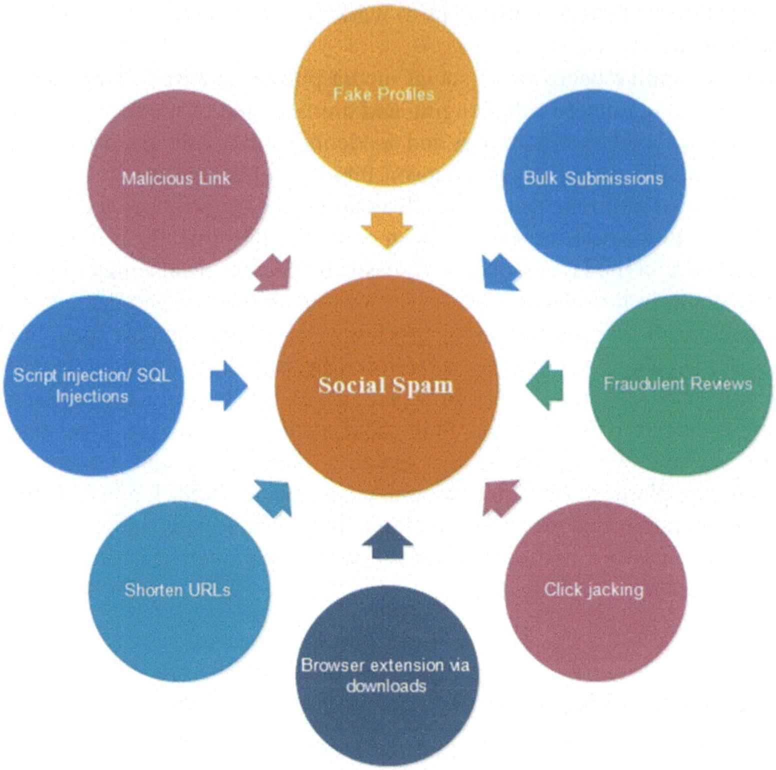

Fig. 11.2 Various categories of social spammer

- Malicious URLs: Malicious URLs damage the potential of the user's account including computer hardware. Some of the malicious URLs spread through a social network (Twitter) are checked by the service provider itself depicted in Fig. 11.3. Malicious URLs spread through various blogs, tweets, posts, direct message services, and many more ways on the social network platform.
- Fake profiles: Adversaries create fake profiles in online social networks to gather confidential information of the user and for some financial benefits. The fake profiles are created on the same platform or on a different platform by collecting user information. The basic objectives of creating fake accounts are to humiliate people over the network and collecting user's credentials from unknown users. Sometimes fake accounts are created to do some fun or some nuisance work.
- Bulk posting or submission: Bulk submission is also called the bombing of bulk spam messages. Through these activities, people attract other users toward their accounts. Sometimes people behave like trustworthy customers and spreads malicious bulk posts over the network. Bulk message contents are similar in nature, i.e., same messages posted many times in equal intervals of time. People

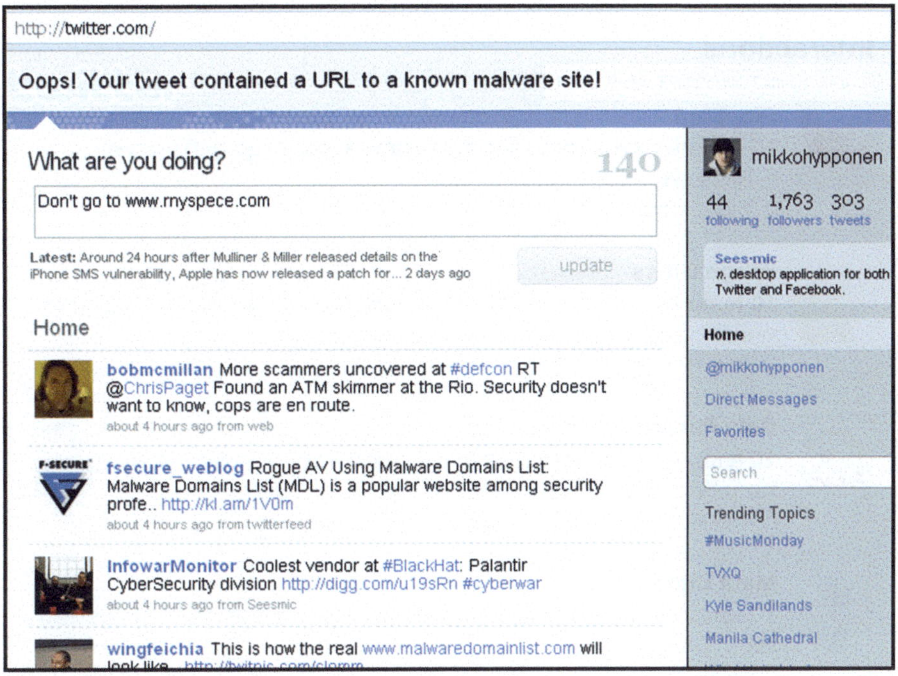

Fig. 11.3 Malicious URLs in Twitter

use certain message spreading tools for spreading bulk posting automatically without user intervention as shown in Fig. 11.4.

- Fraudulent comments: Without knowing the specification of any product and its advantages, people review and claim that the product is good or bad. By this process, the product is highlighted as a good or bad product on social media. With the help of these services, people highlighted their comments and products over the network. Various other forms of social spam include special characters inside comments, harassing news, various threats, and profane words in comments and reviews.
- Spammer through social bot: Bot-based spammer spreading is a new approach by the adversaries in the social network. Bots are created by spammers and spread over the network using message services like Facebook messaging. Some bots are user created and some are system generated. The system-generated bots are spread through certain software.
- Malware-based spammer: Malware is a delivery vehicle for a spammer in a social network platform. Malicious software spread spammers using various tools and services. Some malware is spread through URLs, fraudulent links, and some new approaches in the network.

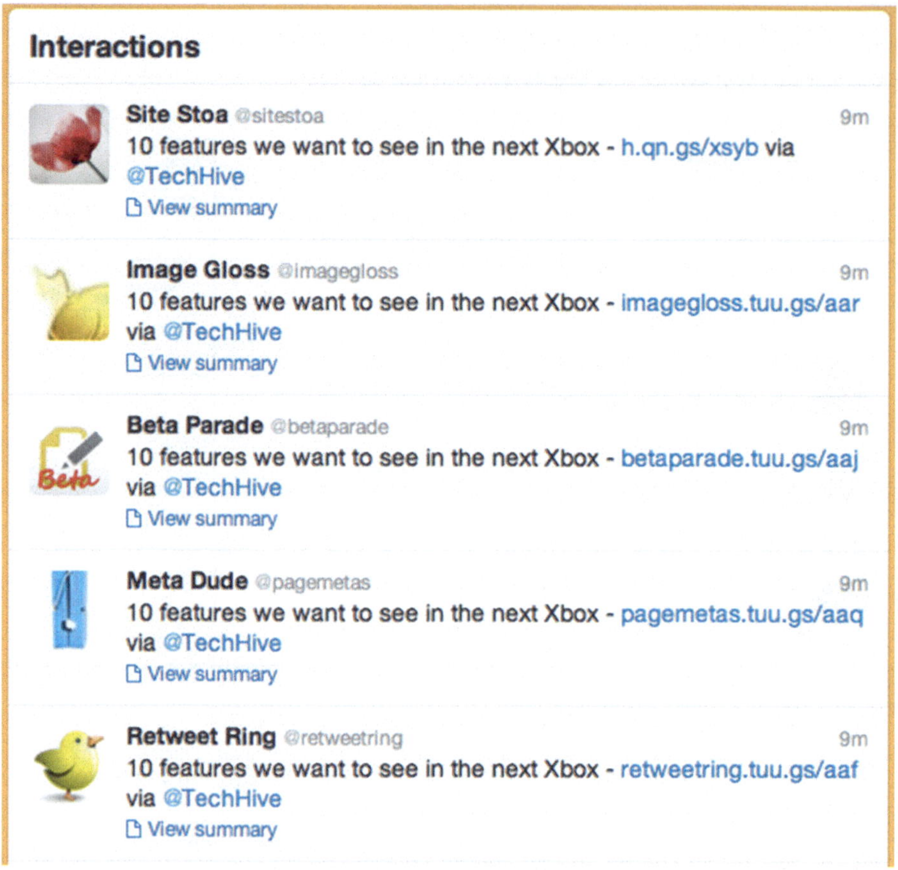

Fig. 11.4 Bulk posting of tweets on Twitter

- Clickjacking: In these spamming techniques, the attacker tries to redirect the users from one page to another by clicking on a link or a blog. When the user visits any blog or clicks on that blog to see the details, the page redirects to some malicious site and malware is downloaded automatically or blocks certain services.
- Update or download malicious browser extension: Malicious software downloaded in the computer via a browser extension. The malicious browser extensions are automatically downloaded without the user's notice and activate some malfunctions in the system. These sorts of services spread through some blogs, reviews with links, advertisements, etc.
- SQL injection: In this type of spammer spreading technique, the user changes the source code of the original content and added some malicious content to behave differently. These techniques spread rapidly over the network in various web

Table 11.1 Types of spam and spreading techniques

	Types of spam								
Spreading technique	Email based	Instant message based	News group based	Search engine based	Blog based	Video based	Social network based	Online game based	Through internet telephony
Malicious link	✓	✓	✓			✓	✓	✓	
Fake profile	✓	✓	✓		✓	✓	✓		
Bulk messaging	✓	✓	✓		✓	✓	✓		
Fraudulent comment/ review	✓	✓	✓	✓	✓	✓	✓		✓
Clickjacking	✓	✓	✓		✓	✓	✓	✓	
Browser extension	✓	✓	✓	✓	✓	✓	✓		
Shorten URLs	✓	✓	✓	✓	✓	✓	✓	✓	
SQL injection (XSS)	✓	✓	✓	✓	✓	✓	✓	✓	
Social bot	✓	✓			✓	✓	✓	✓	✓

applications, social networking sites, Microblogging sites, and review-related blogs. The overview of various spammers and their spreading methods is depicted in Table 11.1.

2.2 *Detection Methods*

The detection of social spam content is difficult to identify due to its hidden nature. Spammer spreads through the social platform over other services like the post, messages, tweets, videos, advertisements, and through direct communications. In [27], authors describe the concept of social spam and state that social spammers are different from other spammers due to their spreading method. They also describe some characteristics of various websites and develop a spam detector model to analyze spam and delete those spam from various websites. The author also identifies and analyzes various combating strategies to detect and identify spam. The authors categorize the spam detection framework into various groups like (1) Identification of spam and removal of the spam content at the same time. (2) Detection of spam and decrease its ranking so that it will not affect the content in the future. (3) Prevention method that protects the user accounts from various threats by blocking spammers. Several approaches based on the above categories are discussed by the authors. Authors in [28] described the concept of spammer detection including their own approach by analyzing temporal evaluation patterns. In this evaluation, pattern authors proposed a dynamic measure to analyze the user's activity and behavior to quantify the user's behavior. Various methods of spam detection by various researchers are described in Sect. 3 with different types and their uses.

3 Literature Review

Classification of the spammer in the social network is a big challenge in the network era due to its epidemic nature. Therefore, to classify the various detection approaches for spammers, we used the analysis of existing categorization including some new ideas depicted in Fig. 11.5. The overall classification of spam detection framework is classified into three different groups called syntax-based, profile feature-based, and blacklisted profile based on uses.

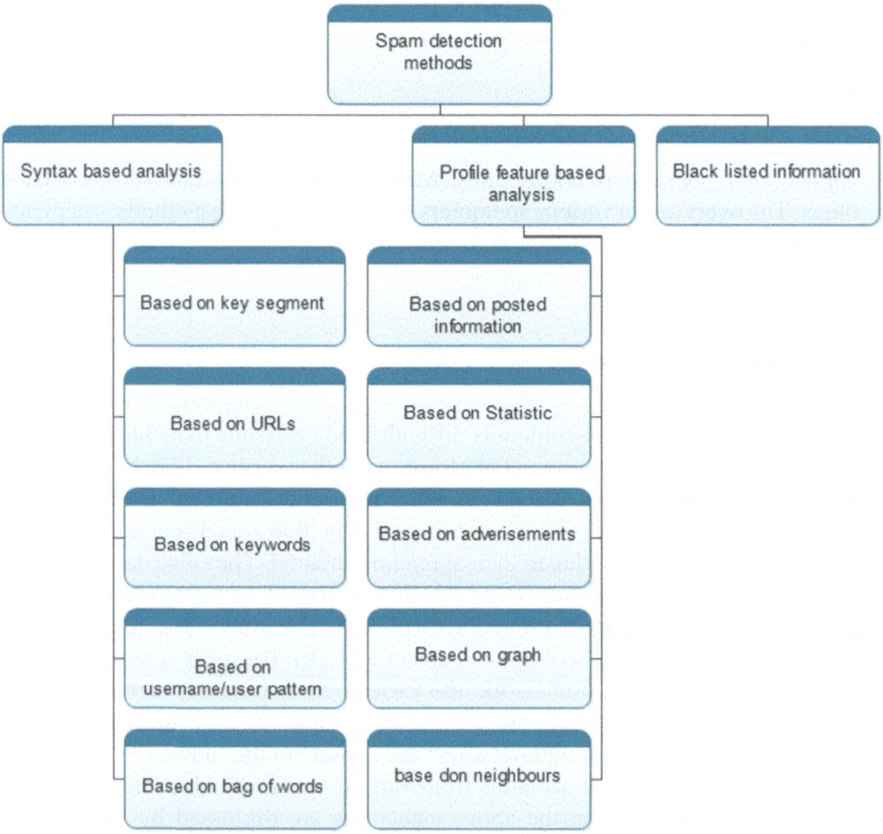

Fig. 11.5 Various spam detection methods

In this section, we analyze various spam detection methods based on various syntax analyses. Various posts are collected from different social platforms based on the shared content and data collection module. By analyzing the post based on the various features, suspicious posts are collected. Using a supervised learning approach, training set and test set modules are created for operations. By analyzing the training set and test set classification, problems are defined and solutions to these problems are generated as spam or non-spam.

3.1 Analysis Based on Key Segments

First, in this literature, we discuss the analysis of spammer detection based on key segments. Various researchers use key segment methods to detect spammers in social network platforms as follows. Detection of spammers by analyzing shorten URLs is the prominent method used by researchers. Some of the Microblogging sites like Twitter limit the number of characters up to 40 for every tweet. To reduce the content into a limited view, various software and algorithms are used by preprocessing the contents [29].

URL-Based Analysis

Some of the Microblogging sites like Twitter limit the number of characters up to 40 for every tweet. To reduce the content into a limited view, various software and algorithms are used for preprocessing the contents [29]. Shortened URLs hide the malicious link that spread spam messages and the original meaning of the content. Therefore, shortening the URLs is a major key segment for analysis. By detecting spammers based on the key segment, the author in [30] develop an identifier. For example, shortened URLs like bit.ly are detected. They calculate the percentage of spammer and non-spammer content used for shortening the URLs. To discriminate between the spammer and non-spammer, they use some division method. If the division value is greater than one, it means the activity is related to the spammer category. Many other methods are also used to duplicate the URLs and classify various posts. Using this method, tweets and accounts related to tweets are clustered into different categories based on the shared URLs. The authors in [27], describe the clustering of the related tweets based on textual content and shortened URLs for better identification of spammers. Based on the URLs shared by the users, the authors in [28] linked all the accounts and formed a cluster for better analysis and identification. After exposure of various Twitter campaigns using URL-based methods, some new algorithms are employed based on a machine learning approach to distinguish spammer contents from regular tweets. All these traditional algorithms are based on statistical feature analysis extracted from different user profiles [28, 31]. The author also uses Shannon's information theory and computes the entropy using the URLs attached in the tweets. Based on the above analysis, the similarity indexed was also calculated and the interval between the tweets computed. Based on the various tweets, a common graph was constructed using the selected similarity index (threshold). Various features are extracted from the URLs like lexical attributes, page content, and domain hosting properties for analyzing spammer activity [32, 33]. The authors in [34] use various features related to IP addresses for analyzing spammer contents. Most of the detection techniques are mixed together for better performance to detect spammer content online. The analysis by the authors in [34, 35] also describes that every analysis of spam based on the URLs included analysis of shortened URLs. Based on the analysis by the author in [30], the likelihood ratio of the shorteners is studied. According to them, they found 77% of spam content accounts in Twitter that are suspended within a single day of their creation. The author in [27] proposed a spam campaign that controlled around 145,000 accounts involved in spreading spam messages in the form of URLs. They collected various features from Twitter accounts like URL redirects, reputation of the account, posting contents, and user account information for their analysis of spam campaigns with the help of Twitter streaming API [36]. We also elaborate on the various advantages and disadvantages of the URL-based analysis. Due to the tweet limit by the service provider, malicious users spread malicious URLs to gather personal information of the user and harass the people over the net. So those contents are included with URLs that should be analyzed for detecting spammer activity in Twitter by various authors. Moreover, the URL-based analysis faced inaccuracy due to shortening of URL features. Various algorithms and reverse engineering principles are applied for changing the shorten URLs to original URLs. In addition to the advantages, various disadvantages are reported by the authors to detect spammer contents based on URL analysis. The main disadvantages related to the URL-based analysis are the fast rate of processing and autorun principle used by the malicious users.

Analysis Based on Pattern or Keyword

To analyze spammer content in tweets, keyword and user name methods also be used by the researchers. But the implementation of this method is very straightforward and intuitive. However, based on our analysis only some researchers work under this category to detect spammer content in social networks in 2009 and 2010. The authors in [37] develop an algorithm to detect spam by detecting keywords and matching the user name. Based on the assumption by the author, the account or the user name combined with letter and numbers have more chance as a spammer account. The tweets spread by the spammer refer to the unsolicited message by manipulating some accounts automatically informal pattern. Also, the author analyzes that the tweets that contain misleading words are more likely from spam accounts. The author in [38] also applied the same principle in 2010 on the Facebook platform. Searching the pattern in the form of malicious content in Facebook and Twitter with the principle of patterns or keywords is a challenging task. The URLs like "click here" are the best example. In practice, both the techniques, i.e., keyword-based and user name-based are applied together for proper analysis of the content and that will also be helpful for shortening URLs. The collection of various information and features are the most important identities for detecting spam content over the social network. Using various social engineering methods, spamming activities in various profiles and social networks can easily avoid the usage of user patterns and keywords.

3.2 Based on Tweet Content

In this section, we discuss the detection of spam content by analyzing various tweet contents posted by users over the network. Because spammer spreads easily over the network using various contents like the bulk of words, fraudulent tweets, and other posted information. We discuss the various methods by which the textual content of tweets can spread. These methods are TF–IDF (Term Frequency–Inverse Document Frequency), Bag of words, and sparse learning and are discussed below.

TF–IDF-Based Analysis

This method of analysis is the most popular technique to eradicate the meaning of various tweets. Various authors who worked in this area are listed as follows. In [39–41], authors use the TF–IDF principle to analyze various tweets to detect malicious contents as spam. Basically, TF–IDF is used to extract text from various posts to identify the context interns of weight [42]. Based on the research in [42], the author in [41] designs a metric to measure the correlation between the tweets in each pair of accounts. The author in [39] also applied the same principle used by the author in [42] to identify the similarity index using the vector space model. By this analysis, the author identifies that the similarity index of legitimate was stronger as compared to spamming one. The TF–IDF-based search first identifies the duplicate tweets posted by the spammer account over the social network platform. The Twitter campaign by the sender is classified into spam and non-spam campaign based on the content reviewed and identified. Based on the vector space model, all the content detected through TF–IDF is processed and evaluated for the best output. As far as performance is concerned, the author in [39] compared eight different machine learning algorithms and found RF (Random Forest) outperformed. As far as features are concerned, the similarity index generated by the TF–IDF technique ranked ninth with an accuracy of 72.3% in random forest classifier. Moreover, the author in [41] combined both tweet content and social relationship and calculated malicious scores between an individual account and its following. The proposed method by the author in [41] called CIA performs better as compared to others and identified 13 more spammers.

Based on Bag of Words

Before training a classifier for classification, the bag of words based method works for the representation of text by preprocessing. Various works have been implemented by various authors based on a bag of words. The method proposed by the authors in [43] using bag of words in TF–IDF techniques as the weighted algorithm to represent vectors. Meanwhile, the basic principle of this algorithm is widely adopted through Bayesian algorithm to pick up words from various paragraphs or posts. Basically, this technique is used to measure statistical analysis like content similarity or classifies the text directly based on their nature or behavior. Feature extractions through bag of word methods are based on text analysis. By this process, various punctuations, lowercasing every character in the sentence or in a tweet, and tokenizing each word can be done through TF–IDF converter converts into texts [43]. For detection of spammer content on tweets and in other posts, these features work individually. Also, these features are associated with other features to work in the detection process of spam. The basic principle of bag of words is also used in Bayesian classifiers called CRM114 [44]. According to the authors in [45], bag of word method is basically used to detect email spam due to simple and easy implementation. In practice, a bag of words combines other features in social network account to detect spammer and their relationship. Simply, only the bag of words method cannot give a suitable solution as like other services. As an example, Bayesian classifier includes tweet descriptions without using any account information.

Based on Sparse Learning

Due to the high-dimensional feature vector generated by the traditional spammer detection method based on n-gram and bag of words, the authors in [46, 47] proposed a sparse representation method. This method represents key phrases or words instead of total sentences. The method was applied by the author in [48] to a non-negative matrix factorization model (NMF). This model is used for the representation of lower dimensional feature vectors. Then, an optimization technique is applied to transfer the text from the next level to the topic level. Due to the shrunk length, features are more representative and make some clusters for better identification of spam. Compared to other models, sparse learning method performs best using the five cross-validation techniques. As a result, this model gives better accuracy.

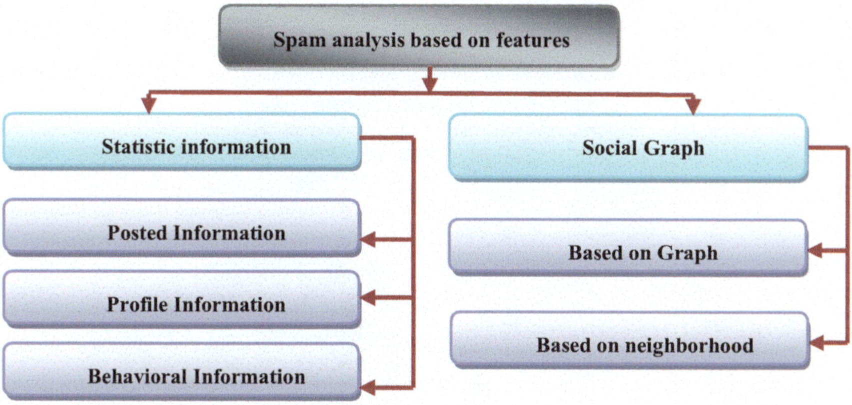

Fig. 11.6 Various categories of features

3.3 *Analysis Based on Features*

In this section, we analyze various spam detection methodologies and frameworks based on the feature analysis. We divide the total features into three different categories called features based on posted information, profile information, and user behavior as depicted in Fig. 11.6. Also, all three categories are coming under broad broad groups.

Based on the Posted Information

Various methods that are based on posts related to social network platforms must work with combined features from all categories for analyzing spammer content in social network platforms. Basically, spammers always spread through social engineering techniques. So the posted information in the social network as text should be analyzed for spammer detection. Information related to the post and its features is depicted in Fig. 11.7.

The author analyzes that the spammer spread more posts as compared to other users in social network platforms [49]. According to the authors in [50], based on cumulative distribution function point, it was reported that spammers usually spread spam content through hashtags, URLs, and spam words within text messages. Also, spammer uses more text size as compared to normal posts [51]. The various features used to detect spammer content in social network platforms are very useful for analysis. Various features are used for spammer detection in social network platforms depicted in the above figure. Based on the statistical feature analysis, it performs well as compared to other methods of feature section. According to the analysis report by the authors in [51], the F-measure score is as high as 93.6% by Random forest classifier. Also, the author analyzes the same using six different classification techniques. Wan et al., 2010 developed a Bayesian network-based method

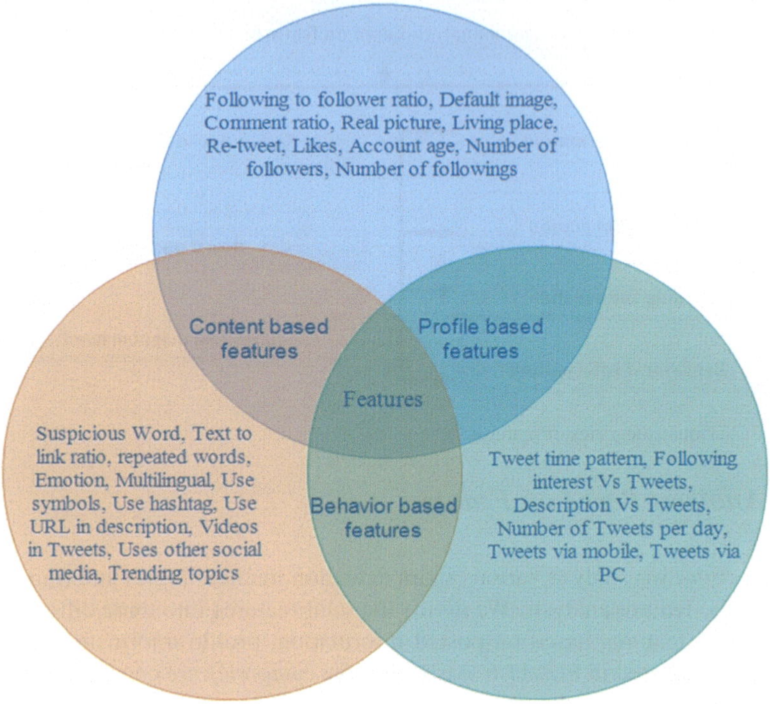

Fig. 11.7 Various features associated with each category

that achieves nearly about 90% TP rate. The author in [52] analyzes the behavior of the account through URL analysis that is embedded in text messages in different social networks like Facebook and Twitter. The authors use decision tree classifiers in [53] using various features to analyze spam content in Facebook. The main objective of the author is to identify the messages that include URLs, hyperlinks, and hashtags. The author in [54] focuses on five various valuable content related to Twitter accounts like content filtering, scalability, proper decision-making, ability to retain the model with new content, and independent of text model developed for detecting spammers in social contents. The most important factor of this model is, it works on the real-time and content filtering option within a short time period. Lee et al. [55] proposed an approach based on various classifications by deploying social honeypots in the network and various user profile features for analysis.

Based on Profile Information

There are also some techniques used by the researchers to identify spammer contents in social network platforms using profile-based feature analysis. According to the authors in [56], the number of followers and followings of spammer account is much larger compared to other users. Also, the author analyzes that the life cycle of spammer users is less as compared to legitimate users in social network platforms. Some of the researchers also analyze spammer content by combining all features into account. According to the authors in [57], the reputation of the spammer is either very low or very high. But the reputation of the legitimate users lies between 30% and 95%. Similar activities found by the authors in [58] were implemented through the Facebook platform. They achieved 98.7% as a true positive rate with 1.4% as a false positive rate. Also, the authors in [50] implemented the same through support vector machine classifier to classify spammer and non-spammer content in Facebook with an 87.2% detection rate. By identifying suspected accounts on Twitter, the author analyzes the activity of the user for detecting spam content [59]. The authors in [60] investigating the deceptive information in Twitter spam by analyzing both public and private information of the user through various features analysis.

Based on the Behavior of the Account

Based on the behavior and campaign in social network platform, various authors analyze the spammer activity. Based on the similarity index generated through feature analysis, authors grouped various spam activities into a cluster of accounts [46–48]. The author in [61] analyzes the behavior of various accounts and the interval between the tweets spread by the user for detecting spammer account. By considering various features and generating content self-similarity scores, the author in [61] analyzes that the spammer embedded some text templates and post similar content over the network. As far as detection rate is concerned, Zhang et al. 2016 measure 88% as F1 score and more than 90% as precision value.

3.4 Analysis Based on Graph

Analysis of spammer content in various social network platforms based on the social graph is a major challenge by the researchers. The overall detection method is implemented using various features associated with the follower and following activities of the user. Based on the follower and following of user graph-based method categorized into two different groups called graph-based method and neighborhood methods. Each node in the social graph-based method represents an account with in-degree and out-degree nodes. The in-degree of a node denotes followers and the out-degree node is called followings [46, 47]. The authors in [47] analyze the behavior of the account through graph structures and the features associated with the account, i.e., graph density, reciprocity, and shortest path. Another researcher in [62] analyzes the BC (Betweenness centrality) values to identify the activity and association of the account. In recent years, neighborhood methods for detecting social network spam is the main concern. Based on this method, various features associated with the accounts are collected for spammer analysis. According to the author [62], neighbor accounts are affected by the spammer over social network platforms. The information related to the posted content can be identified by the follower or followings account.

4 Comparative Analysis

Comparative analysis of various methods used by the authors for detecting spammer content in social network platforms was depicted in Table 11.2.

5 Open Issues and Challenges

In this survey, we have discussed various methods and techniques for detecting spammer content in social network platforms. As we can see from various methods, the majority of the analysis mainly depends on the machine learning platform. Among all these techniques, the major differences are identified based on the method and feature selections. Our literature survey reviewed various methods and techniques for identifying spammer content in social network platforms. Also, there are several open issues and challenges for existing methods. We identified and present some of the open issues in this section. First, the collection of the real dataset is a challenging task. Real-time datasets are required for better analysis of spammers. Second, labeling the dataset manually is too difficult. So, the proper methodology should be applied for labeling raw data into the labeled dataset. Third, both public and private information are required for better analysis of spammer and account related to spammer category. Fourth, proper classification techniques are applied for better decision-making. Finally, fabrication of the data is used to train and test the appropriate model and is easy to manipulate from time to time.

6 Conclusion

In this chapter, we review the various state of arts related to spammer detection in social network platforms. We first categorize the type of spam spread through the social network by the spammer user. We further carried out the spammer detection techniques with thepros and cons of every method and also the comparative analysis of existing approaches. It was found that the spammers are spread through social

Table 11.2 Comparative analysis of various spam detection framework based on their features used

Authors	Title	Technique used	Feature analysis	Accuracy (%)	Pros and cons
Aslan et al. [63]	Automatic detection of cybersecurity-related accounts on online social networks	Machine learning-based classifiers with prototypical words	User-based and behavioral features for analysis	97.17	Better accuracy compared to other methods
Sohrabi et al. [64]	A feature selection approach to detect spam in the Facebook social network	PSO-based hybrid method for spam analysis	Optimization-based feature selection	91.20	Extraction of features from various profiles but less number of features for analysis
Singh et al., 2018 [65]	Who is who on twitter—Spammer, fake, or compromised account? A tool to reveal true identity in real time	Various machine learning approaches with feature selection methods	Feature related to pornographic contents	92.1	Less number of feature selection with manual selection process
Erwin et al. [66]	Detecting Indonesian spammer on Twitter	SVM and other machine learning approaches	User behavior and post content features	93.67	Manual selection of features but detection rate measures better performance in terms of accuracy
Bindu et al. [67]	Discovering spammer communities on Twitter	Graph-based approaches including machine learning-based analyzer	Community-based features including structural characteristics	86.7	Cluster-based approach need more features for analysis
Gupta et al. [4]	Collective classification of spam campaigners on Twitter: A hierarchical meta-path-based approach	Hierarchical meta-path detection mechanism for analysis	User-based and behavioral-based features for analysis	67.3	Detection rate is very low but hierarchical method gives better direction for detection of spam

(continued)

Table 11.2 (continued)

Authors	Title	Technique used	Feature analysis	Accuracy (%)	Pros and cons
Chu et al. [27]	Detecting social spam campaigns on Twitter. In: International conference on applied cryptography and network security	Clustering the related tweets based on textual content and shortened URLs for better identification of spammers	Features related to shortening the URLs and textual content features	87	URLs cannot be detected for spam content with better accuracy
Yardi et al. [36]	Detecting spam in a Twitter network	Analysis based on patterns and keywords	User behavior-based features for spam content analysis in Twitter	91	Pattern-based detection system is appropriate but the keyword-based method is not suitable for larger dataset
Benevenuto et al. [50]	Detecting spammers on Twitter. In: Collaboration, electronic messaging, anti-abuse, and spam conference	Cumulative distribution function point was reported that spammers usually spread spam content through hashtags and URLs	Features related to URLs and hashtags	–	Only URL-based features cannot be detected for spam content with better accuracy
Chen et al. [51]	Six million spam tweets: A large ground truth for timely Twitter spam detection	Random forest-based classification in machine learning environment	Features related to the Tweets and posted information, i.e., content-based features	93.6	Only content-based features are not sufficient for analysis
Ahmed et al. [58]	A generic statistical approach for spam detection in online social networks	Finding reputation of the spammer as either very low or very high. But, the reputation of the legitimate users lies between 30% and 95%	User profile features are used for analysis	98.7	Reputation of the user cannot be identified through profile features analysis

(continued)

Table 11.2 (continued)

Authors	Title	Technique used	Feature analysis	Accuracy (%)	Pros and cons
Zhang et al. [61]	Detecting spam and promoting campaigns in Twitter	Author analyzes that the spammer embedded some text templates and posted similar content over the network	User account and posted information-based features are analyzed	90	Less number of feature selection with manual selection process leads to lower accuracy and detection rate

network contents and that can be detected through various methodologies including futures related to user account and post. Finally, we made a brief summary and discussed some open issues related to social network spam detection. We hope this survey helps a lot to the researchers and the users who participated in the networks for sharing information like Facebook, Twitter, Instagram, etc.

References

1. Gupta BB (ed) (2018) Computer and cyber security: principles, algorithm, applications, and perspectives. CRC Press, Boca Raton
2. Fire M, Goldschmidt R, Elovici Y (2014) Online social networks: threats and solutions. IEEE Commun Surv Tutor 16(4):2019–2036
3. Sahoo SR, Gupta BB (2019) Classification of multiple attacks and their defense mechanism in online social networks: a survey. Enterp Inf Syst 13(6):832–864
4. Ho K, Liesaputra V, Yongchareon S, Mohaghegh M (2018) Evaluating social spammer detection systems. In: Proceedings of the Australasian computer science week multiconference, January. ACM, p 18
5. Gupta, S., Khattar, A., Gogia, A., Kumaraguru, P., & Chakraborty, T. (2018). Collective classification of spam campaigners on twitter: a hierarchical meta-path based approach. arXiv preprint arXiv:1802.04168
6. Stergiou CL, Psannis KE et al (2020) IoT-based big data secure management in the fog over a 6G wireless network. IEEE Internet Things J 8:5164–5171
7. Mishra A, Gupta N, Gupta BB (2021) Defense mechanisms against DDoS attack based on entropy in SDN-cloud using POX controller. Telecommun Syst 77:1–16
8. Alsmirat MA, Al-Alem F, Al-Ayyoub M, Jararweh Y et al (2019) Impact of digital fingerprint image quality on the fingerprint recognition accuracy. Multimed Tools Appl 78(3):3649–3688
9. Dahiya A, Gupta BB (2021) A reputation score policy and Bayesian game theory based incentivized mechanism for DDoS attacks mitigation and cyber defense. Future Gener Comput Syst 117:193–204
10. Bhushan K, Gupta BB (2019) Distributed denial of service (DDoS) attack mitigation in software defined network (SDN)-based cloud computing environment. J Ambient Intell Humaniz Comput 10(5):1985–1997
11. Olakanmi OO, Dada A (2019) An efficient privacy-preserving approach for secure verifiable outsourced computing on untrusted platforms. Int J Cloud Appl Comput 9(2):79–98

12. Hossain MS, Muhammad G, Abdul W, Song et al (2018) Cloud-assisted secure video transmission and sharing framework for smart cities. Futur Gener Comput Syst 83:596–606
13. Kaushik S, Gandhi C (2019) Ensure hierarchal identity based data security in cloud environment. Int J Cloud Appl Comput 9(4):21–36
14. Gou Z, Yamaguchi S (2017) Analysis of various security issues and challenges in cloud computing environment: a survey. In: Identity theft: breakthroughs in research and practice. IGI Global, Hershey, pp 221–247
15. Cvitić, I., Peraković, D., Periša, M., & Botica, M. (2021). Novel approach for detection of IoT generated DDoS traffic. Wireless Networks, 27(3), 1573–1586
16. Cvitic I, Perakovic D, Perisa M, Botica M (2020) Definition of the IoT device classes based on network traffic flow features. In: Knapcikova L, Balog M, Perakovic D, Perisa M (eds) EAI/Springer innovations in communication and computing [internet]. Springer, Cham, pp 1–17
17. Perakovic D, Perisa M, Cvitic I, Husnjak S (2017) Artificial neuron network implementation in detection and classification of DDoS traffic. TELFOR J 9(1):26–31
18. Pasupuleti SK (2019) Privacy-preserving public auditing and data dynamics for secure cloud storage based on exact regenerated code. Int J Cloud Appl Comput 9(4):1–20
19. Al-Qerem A, Alauthman M, Almomani A et al (2020) IoT transaction processing through cooperative concurrency control on fog–cloud computing environment. Soft Comput 24(8):5695–5711
20. Cvitić I, Peraković D, Periša M, & Jurcut AD (2021) Methodology for Detecting Cyber Intrusions in e- Learning Systems during COVID-19 Pandemic. Mobile networks and applications, 1–12
21. Cvitić I, Peraković D, Periša M, Husnjak S (2019) An overview of distributed denial of service traffic detection approaches. Promet Traffic Traffico 31(4):453–464
22. Cvitić I, Peraković D, Periša M, Gupta BB (2021) Ensemble machine learning approach for classification of IoT devices in smart home. Int J Mach Learn Cybern Ensemble 12:1–24
23. Ahmed H (2017) Detecting opinion spam and fake news using n-gram study and semantic similarity. Ph.D. thesis
24. Sahoo SR, Gupta BB (2019) Hybrid approach for detection of malicious profiles in twitter. Comput Elect Eng 76:65–81
25. Yao Y, Viswanath B, Cryan J, Zheng H, Zhao BY (2017) Automated crowdturfing attacks and defenses in online review systems. In: Proceedings of the ACM SIGSAC conference on computer and communications security (CCS), Dallas, TX, USA, pp 1143–1158
26. Sahoo SR, Gupta BB (2020) Multiple features based approach for automatic fake news detection on social networks using deep learning. Appl Soft Comput 100:106983
27. Sahoo SR, Gupta BB (2020) Real-time detection of fake account in twitter using machine-learning approach. In: Advances in computational intelligence and communication technology. Springer, Singapore, pp 149–159
28. Sahoo SR, Gupta BB, Choi C, Hsu CH, Chui KT (2020) Behavioral analysis to detect social spammer in online social networks (OSNs). In: International conference on computational data and social networks. Springer, Cham, pp 321–332
29. Klien F, Strohmaier M (2012) Short links under attack: geographical analysis of spam in a url shortener network. In: Proceedings of the 23rd ACM conference on hypertext and social media. ACM, pp 83–89
30. Thomas K, Grier C, Ma J, Paxson V, Song D (2011) Design and evaluation of a real-time url spam filtering service. In: 2011 IEEE symposium on security and privacy. IEEE, pp 447–462
31. Zhang X, Zhu S, Liang W (2012) Detecting spam and promoting campaigns in the twitter social network. In: 2012 IEEE 12th international conference on data mining. IEEE, pp 1194–1199
32. Ma J, Saul LK, Savage S, Voelker GM (2009) Identifying suspicious urls: an application of large-scale online learning. In: Proceedings of the 26th annual international conference on machine learning. ACM, pp 681–688
33. Whittaker C, Ryner B, Nazif M (2010) Large-scale automatic classification of phishing pages. In: NDSS, vol 10

34. Sahoo SR, Gupta BB (2020) Classification of spammer and nonspammer content in online social network using genetic algorithm-based feature selection. Enterp Inf Syst 14(5):710–736
35. Sahoo SR, Gupta B, Choi C, Esposito C (2020) Detection of spammer account through rumor analysis in online social networks. In: The 9th international conference on smart media and applications. (pp. n-a)
36. Twitter Developers. Twitter's streaming API documentation; 2016. Available from: https://dev.twitter.com/streaming. Accessed 23 June 2019
37. Yardi S, Romero D, Schoenebeck G et al (2009) Detecting spam in a twitter network. First Monday 15(1). https://doi.org/10.5210/fm.v15i1.2793
38. Gao H, Hu J, Wilson C, Li Z, Chen Y, Zhao BY (2010) Detecting and characterizing social spam campaigns. In: Proceedings of the 10th ACM SIGCOMM conference on Internet measurement. ACM, pp 35–47
39. Sahoo SR, Gupta BB (2020) Fake profile detection in multimedia big data on online social networks. Int J Inf Comput Secur 12(2–3):303–331
40. Ivan Cvitić, G. Praneeth, D. Peraković (2021), Digital Forensics Techniques for Social Media Networking. Insights2Techinfo, pp.1
41. Yang C, Harkreader R, Zhang J, Shin S, Gu G (2012) Analyzing spammers' social networks for fun and profit: a case study of cyber criminal ecosystem on twitter. In: Proceedings of the 21st international conference on world wide web. ACM, pp 71–80
42. Salton G, Buckley C (1988) Term-weighting approaches in automatic text retrieval. Inf Process Manage 24(5):513–523
43. Lee K, Caverlee J, Webb S (2010) Uncovering social spammers: social honeypots+ machine learning. In: Proceedings of the 33rd international ACM SIGIR conference on research and development in information retrieval. ACM, pp 435–442
44. Gupta S, Gupta BB, & Chaudhary P (2018) Hunting for DOM-Based XSS vulnerabilities in mobile cloudbased online social network. Future Generation Computer Systems, 79, 319–336
45. Chaudhary P, Gupta BB, & Gupta S (2019) A framework for preserving the privacy of online users against XSS worms on online social network. International Journal of Information Technology and Web Engineering (IJITWE), 14(1), 85–111
46. Khushboo Kumari (2021) Online social media threat and It's solution, Insights2Techinfo, pp.1
47. Hu X, Tang J, Liu H (2014) Online social spammer detection. In: AAAI. ACM, New York, pp 59–65
48. Lee DD, Seung HS (1999) Learning the parts of objects by non-negative matrix factorization. Nature 401(6755):788–791
49. Yang C, Harkreader RC, Gu G (2011) Die free or live hard? Empirical evaluation and new design for fighting evolving twitter spammers. In: International workshop on recent advances in intrusion detection. Springer, Cham, pp 318–337
50. Benevenuto F, Magno G, Rodrigues T, Almeida V (2010) Detecting spammers on twitter. In: Collaboration, electronic messaging, anti-abuse and spam conference (CEAS), vol 6, p 12
51. Chen C, Zhang J, Chen X, Xiang Y, Zhou W (2015) 6 million spam tweets: a large ground truth for timely twitter spam detection. In: 2015 IEEE international conference on communications (ICC). IEEE, pp 7065–7070
52. Cao C, Caverlee J (2015) Detecting spam URLs in social media via behavioral analysis. In: Proceedings of advances in information retrieval. Springer, pp 703–714
53. Soiraya M, Thanalerdmongkol S, Chantrapornchai C (2012) Using a data mining approach: spam detection on Facebook. Int J Comput Appl 58(13):26–31
54. Thomas K, Grier C, Ma J, Paxson V, Song D (2011) Design and evaluation of a real-time url spam filtering service. In: Proceeding of IEEE symposium on security and privacy (SP)
55. Lee K, Caverlee J, Webb S (2010) Uncovering social spammers: social honeypots + machine learning. In: Proceedings of the 33rd international ACM SIGIR conference on research and development in information retrieval, SIGIR '10, pp 435–442
56. Chen C, Wen S, Zhang J, Xiang Y, Oliver J, Alelaiwi A et al (2017) Investigating the deceptive information in twitter spam. Futur Gener Comput Syst 72:319–326

57. Wu T, Liu S, Zhang J, Xiang Y (2017) Twitter spam detection based on deep learning. In: Proceedings of the Australasian computer science week multiconference. ACM, p 3
58. Ahmed F, Abulaish M (2013) A generic statistical approach for spam detection in online social networks. Comput Commun 36(10):1120–1129
59. Thomas K, Grier C, Song D, Paxson V (2011) Suspended accounts in retrospect: an analysis of twitter spam. In: Proceedings of the 2011 ACM SIGCOMM conference on internet measurement conference, IMC '11, pp 243–258
60. Chen C, Wen S, Zhang J, Xiang Y, Oliver J, Alelaiwi A, Hassan MM (2017) Investigating the deceptive information in twitter spam. Futur Gener Comput Syst 72:319–326
61. Zhang X, Li Z, Zhu S, Liang W (2016) Detecting spam and promoting campaigns in twitter. ACM Trans Web 10(1):4
62. Yang Z, Wilson C, Wang X, Gao T, Zhao BY, Dai Y (2014) Uncovering social network Sybils in the wild. ACM Trans Knowl Discov Data 8(1):2
63. Yang C, Harkreader R, Gu G (2013) Empirical evaluation and new design for fighting evolving twitter spammers. IEEE Trans Inf Forensics Secur 8(8):1280–1293
64. Aslan, Ç. B., Sağlam, R. B., & Li, S. (2018, July). Automatic detection of cyber security related accounts on online social networks: Twitter as an example. In Proceedings of the 9th International Conference on Social Media and Society (pp. 236–240).
65. Sohrabi MK, Karimi F (2018) A feature selection approach to detect spam in the Facebook social network. Arab J Sci Eng 43(2):949–958
66. Singh M, Bansal D, Sofat S (2018) Who is who on Twitter–spammer, fake or compromised account? A tool to reveal true identity in real-time. Cybern Syst 49(1):1–25
67. Setiawan EB, Widyantoro DH, Surendro K (2018) Detecting Indonesian spammer on Twitter. In: 2018 6th international conference on information and communication technology (ICoICT), May. IEEE, pp 259–263

Chapter 12
Overview of Wi-Fi-Based Automatic Passenger Counting Solutions in Public Urban Transport

Ivan Grgurević , Karlo Juršić, and Vinko Rajič

1 Introduction

In the twentieth and twenty-first centuries, the major societal change that characterized those periods was the trend of urbanization (the displacement of population from rural to urban environment). This process created a new demand for transportation and generated new ideas such as urban mobility, people mobility analytics, Smart Cities, and many others. Urban mobility is increasing proportionally to the development of the urban environment, but in some parts of the world it is even faster and developing exponentially. This is caused by an increased number of people and citizens in certain urban environments that sometimes generate faster than the city infrastructure can follow. This increasing number of people in the cities creates an ever-increasing demand for faster, cheaper, more comfortable, safer, and more ecologically conscious urban transport. And so, public transport is the primary mode of transport for a large number of people worldwide. However, there is still a large proportion of the global population that uses passenger cars as their primary mode of transportation. For example, most people in the European Union use their personal cars as their primary mode of transportation (4 trillion passenger-kilometers per year) and bus and tram/metro are less than an eighth of that (569 billion passenger-kilometers for the bus and 89 billion for tram/metro) [1]. This means that car traffic accounted for 71% of the total share of passenger mileage in the EU for 2018, while the bus occupied 8% and the tram/metro accounted for 1.4% [2].

In response to that, increasing development and further optimization processes for PPT are designed and created (alongside urban housing, business and

I. Grgurević (✉) · K. Juršić · V. Rajič
Faculty of Transport and Traffic Sciences, Department of Information and Communications Traffic, University of Zagreb, Zagreb, Croatia
e-mail: igrgurevic@fpz.unizg.hr

L. Knapcikova et al. (eds.), *Sustainable Management of Manufacturing Systems in Industry 4.0*, EAI/Springer Innovations in Communication and Computing,
https://doi.org/10.1007/978-3-030-90462-3_12

commercial hubs, and districts) as an important component in the expansion of cities and urban environments. It can be argued that transportation is the most important link in the urban environment chain and its proper operation is of the utmost importance for the city to function. In the future, the ratio of using personal cars and PPT vehicles will have to change significantly if there is going to be an economical, ecological, and sustainable future, making the Smart City concept a reality and for the transition into the Society 5.0. Even so, the study [3] from 2014 shows that the largest percentage of passengers in the European Union, 59% of them, use some form of PPT—tram, bus, metro, etc.—in their everyday life [3].

To decrease the usage of personal transportation in favor of the public transportation, one needs to understand the problems of urban city traffic. For this to happen, the officials in charge need to understand that dealing with the increasing traffic and traffic congestion does not necessarily mean increasing the infrastructure capacities, but can mean transferring a part of the people from personal transportation to public transportation, if only to commute (peak rush hours). The best way to accomplish this is to optimize the public urban transport network so it is a viable option over personal transportation. Creating a viable alternative means having a great understanding of the PPT network and people's mobility to design and improve the system in real time. One of the best ways to increase the number of passengers who use PPT is to optimize the experience and travel times and the best way to do so is to use Automatic Passenger Counting (APC) to gather data in real time.

Any form of counting transport and transported entities should be the backbone of traffic and transport and the information gathered is used to design, create, reconstruct, and optimize the transport network or infrastructure. The counting data can be used to notice certain patterns in behavior for certain locations and time periods which can then be used to redistribute resources and personnel to where they are needed in accordance with the overall goals and purposes. The correct assessment or a precise number of passengers in PPT is of significance for proper management and organization of public urban transport. There are many different methods of counting passengers in public transport, as shown in paper [4], but this paper has focused on the Wi-Fi-based ones that can vary in method, mounting locations (inside the vehicle or in the station), and can provide various data to be used for different purposes (ridership, OD matrix, driving schedule, any form of people mobility analytics—PmA, etc.). Wi-Fi is not the only wireless technology (Bluetooth, ZigBee, etc.) that can be used for this purpose, but it has been chosen because many transport service providers already have some form of Wi-Fi capability in their PPT vehicles. Also, a great advantage of APC solutions based on Wi-Fi is the increasing number of smartphone users worldwide (from 1 billion in 2012 to 3.8 billion in 2021) [5].

The goal of the research was to provide an overview of automatic passenger counting solutions based on Wi-Fi in public city transport, i.e., to determine different approaches to passenger counting using Wi-Fi technology. The motive for researching these topics is related to the unavailability of the APC taxonomy, i.e., the clear division and applicability of the method of counting passengers using Wi-Fi technology in various subsystems of public urban transport, as well as the widespread usage of integrated Wi-Fi networks in PPT vehicles worldwide. While investigating the problem of counting passengers via Wi-Fi, the problem of inconsistency of terminology related to active and passive counting methods was noticed, which is described further in the paper.

The second section provides an overview and review of the existing literature and research in the analyzed area. The third section describes the features of Wi-Fi as the basis for APC solutions. Furthermore, different scanning methods that represent the terminology problem are described and reviewed in the context of passenger counting. In the fourth section, analysis of the results of existing research is presented according to several different factors, such as the location of the counting (in the vehicle/outside the vehicle), the observed means of transport (bus, train, metro, etc.), the purpose of the collected data, estimation, real counted values, and methods/methodology of passenger counting. It is explained why some information was separated from reviews and what are the major connections that link most of the conducted experiments. In the fifth section, all the findings and conclusions of the research were synthesized and guidelines and plans for future research were given.

2 Existing Research

The authors of paper [6] researched and experimented with the concept of passenger counting based on Wi-Fi sensing system (media access control or MAC address collecting) and signal strength analysis with camera-based APC for reviewing and checking results. Their research was conducted at four bus stops along the main transit corridor in Charlottesville, USA. The results they gathered proved that their solution has a precision of 91% for bus passengers and that the waiting time had an error rate of only 7 seconds. They concluded that their system can be a viable and low-cost solution with high potential with Wi-Fi sensing. Paper [7] has proposed statistical methods for estimating OD matrices for public transport trip level based on collected Wi-Fi data. The authors have installed their equipment in 32 buses in the city of Stavanger in Norway and have collected Wi-Fi data (probe request frames and time between the first and the last frame) for a period of several months and based on those data they have analyzed the probability for each detected device to be on or off the bus and estimated the OD matrix. There was no base truth data for the OD matrix so some small errors in estimates are possible. Using data from collecting probe request frames is the basis of research in paper [8] and the case study was conducted in the city of Obuse in Japan inside a "hop-on-hop-off" type of bus. The goal was to estimate bus passenger volume using a Wi-Fi scanner that collected MAC addresses and global positioning system (GPS) data and clean the raw data enough to perform the estimation effectively. The authors have concluded that the correlation between real values and their estimation is high and amounts to 0.78. They continue to describe the process for collecting and processing said data to generate the OD matrix and speed calculations to determine if the MAC address timestamp data and GPS logs correlate to the actual bus speed. The problems they voice are the number of Wi-Fi-capable devices that are almost always a little wrong because some passengers carry more than one device and some might not have any and some devices can change their MAC address after a software upgrade. Another limitation they encountered was the large range Wi-Fi sensors have and so, a lot of noise is generated in the collected datasets. This can of course be countered by using a directional antenna or by lowering the effective covering range.

Authors of paper [9] state that their solution is low-cost infrastructure able to collect geolocation data and provide services with the goal of optimizing the routes/lines and driving schedules by providing real-time information. Their devised system was deployed in 19 PPT vehicles (buses), on the Madeira Island in Portugal, to collect Wi-Fi data which they then analyzed and compared to, what they considered, ground truth data gathered from ticket validation. They also used the collected data to estimate the number of passengers that exit/enter, detect peak values, and OD matrices. The idea, again, is to use the existing Wi-Fi infrastructure to collect MAC addresses and geolocation data, processed to exclude random MAC addresses by filtering those that come up just once per trip, for mapping bus routes and stops and to log bus trips. The drawback of the proposed solution is the passive nature of collecting probe frames that is dependent on the AP's frequency. However, this solution can be adapted for monitoring crowd levels, disruptions, alerts, waiting times and lines, etc. The value of Wi-Fi data and smartphone devices in everyday life is also shown in paper [10]. The authors attempt to measure the correlation between counting Wi-Fi probe request frames and the real number of people. Also, they compare their results with manual counting results conducted for 8 hours and conclude that the results are promising and the correlation is high (0.89322). They also conclude that the used method is adequate for estimating the number of people, even though it cannot reliably provide the exact number and that it can be used to calculate and estimate the waiting times in correlation with passengers joining or leaving the crowd. However, the authors have not conducted their research in public transport vehicles or stations, but rather in a large hall with a large crowd of people.

Paper [11] describes a solution with Wi-Fi sensor logs from smartphone devices and proposes an unsupervised "Passively Encountering Group" (PEG) detection framework. Every smartphone in the vicinity of Wi-Fi AP is detected and the signal strength is measured and logged. The log is then sent to the server where the information is used to form PEGs. The developed model identifies the subjects in the proximity of the AP based on the strength of the Wi-Fi signal and then has a score attached to it based on their metric system to determine the level of proximity. The experiment was conducted on the university campus of the Indian Institute of Technology in the city of Kharagpur and included over 25 subjects for a time period of 6 months. The conclusion was that the precision for this solution is higher than 90% and the authors state that they have identified two problems in their experiment, which are identification of subjects in the proximity and detecting groups from the proximity information. Another research experimenting with signal strength in Wi-Fi networks 12 has a proposed solution for estimating the number of people based on a regression-based system [12]. The research was conducted in a controlled environment in the Kindai University laboratory in Osaka using a Wi-Fi

AP and a node (Raspberry Pi 2 Model B with a USB Wi-Fi module). The authors state that the precision is higher using a support vector regression-based system rather than the linear regression-based one. They also conclude that the precision of the support vector regression-based system for estimating people is 0.772, for the degree of congestion it is 0.946, and for the presence or absence of people it is 0.982, but they state that they need to evaluate the proposed solution in different environments such as large halls, outdoors, and other similar places, as well as to increase the number of used APs.

The authors of paper [13] describe and analyze their proposed solution for monitoring and predicting the movements of people using, what they call, a passive Wi-Fi sensing system. However, their experiment was also performed in enclosed locations, in two different offices in Singapore, and the authors state that this is because of the open space and dynamic environments. The main proposition is the estimation of the Length Of Stay (LOS) in real time and is based on collecting probe requests and other types of frames. They use a linear Support Vector Machine (SVM) classifier and an online Stochastic Gradient Descent (SGD) mechanism to account for the dynamic environment. The conclusion of this experiment is that SGD is optimal for dynamic and rapidly changing environments and SVM is more precise in less dynamic ones. The results from this are SVM—no more than three seconds—and SGD—less than one second to train on average. The solution proposed by the authors of paper [14] is based on estimating the passenger occupancy and flow in buses using a mobile application capable of detecting other mobile devices over Bluetooth and Wi-Fi. They state that, using a binary classification model, they can determine whether a bus is full or not based on the number of Wi-Fi users. The application continuously scans for devices and counts how many it has detected and then sends the data to the server to be processed. The experiment was performed in the city of Dublin in Ireland, with volunteers who agreed to install and use their application for one week. They would then activate the application once on the bus and would count the real value to submit to the system. Afterward, the authors developed their model and concluded that it has an almost 88% precision in the performed tests. Also, the authors of paper [15] proposed a feasibility study to estimate the bus passengers and congestion they generate by detecting Wi-Fi probe requests in buses and at bus stops. The experiment was conducted at the bus stop in the Ritsumeikan University in Japan and on board the bus at the local train station. The data were collected at the bus stops as well as on board buses during peak hours. The analysis performed using the Pearson correlation showed the correlation of 0.6215 and constant of proportionality, devices/passengers, around 5.5. They state that congestion might also be estimated but the collected data were insufficient. This method might also be used to show trends in behavior of passengers and their daily routine. One more research that proposes solutions for estimating the number of passengers/Wi-Fi devices via probe request collecting is paper [16]. The stated purpose of the experiment was to avoid the problem of random MAC addresses using the propagated information with no need for the real MAC address. The paper was a continuation of an already conducted experiment that had analyzed the probe request process in detail [17]. The proposed solution was modeled to detect Wi-Fi devices when they enter/exit and while they are in the detection zone and the experiment was conducted in the laboratory conditions at the Federal University of Rio de Janeiro. The analyzed solution has shown high correlations (Herlock 0.087, LR-based method 0.383, SVR-based method 0.298) between the number of mobile devices and the real number of people/passengers. Also, Pearson's correlation calculation has shown high correlation as well—0.896—with a low mean relative error—0.087.

3 Features of Wi-Fi as the Basis of an APC Solution

The most reliable information on the actual number of passengers in vehicles is obtained based on the number of arrivals and departures of passengers at each station along the public transport line. Based on this, highly reliable information can be collected and then used to determine the distribution of trip times and efficiency of individual lines and vehicles, optimize the timetables, analyze the vehicle routes and lines, increase or shorten routes, add or remove certain stations, etc. However, conducting such experiments requires significant financial investment, as well as other resources (human and time), so it is necessary to harmonize data collection in accordance with the capabilities of the subjects requesting the measuring. Traffic counting, i.e., passenger counting is carried out by manual or automatic methods and the choice primarily depends on the type of information collected, the duration of the count, and the available financial budget. Although manual counting methods are still used, efforts are being made to automate passenger counting to reduce the need for human components in the process. The reason is the unreliable and unpredictable nature of the people in the counting process, given that it is subject to human error (fatigue, distraction, forgetfulness, etc.).

It is the automatic counting of passengers that eliminates "subjective" circumstances, which in turn eliminates the problem of the human error factor that affects the collection of data. There are currently a number of commercial and "improvised" automatic counting solutions available for the implementation as seen in the existing research that are further explained in Sect. 4. Nevertheless, globally, their application in PPT vehicles is still not widely applied and the reasons are multiple, but boil down to the fact that a large number of PPT service providers do not have the financial power to procure automated counters and there is no universal solution that is applicable in all situations. The absence of universal solutions means different public service providers use different means of transport that operate in different, variable conditions (different vehicle dimensions and types, unpredictability of traffic in urban areas, etc.). Still, pilot experiments with commercial and "improvised" solutions are being conducted with increasing interest to find an efficient and reliable way to collect passenger data. Furthermore, the main feature of the APC solutions is manifested in the fact that they are able to respond to the imposed demands caused precisely by the unpredictability and dynamic nature of urban traffic as well as any other environment in which they operate. Also, when passenger counting is performed, it is unnecessary to implement the APC components in all

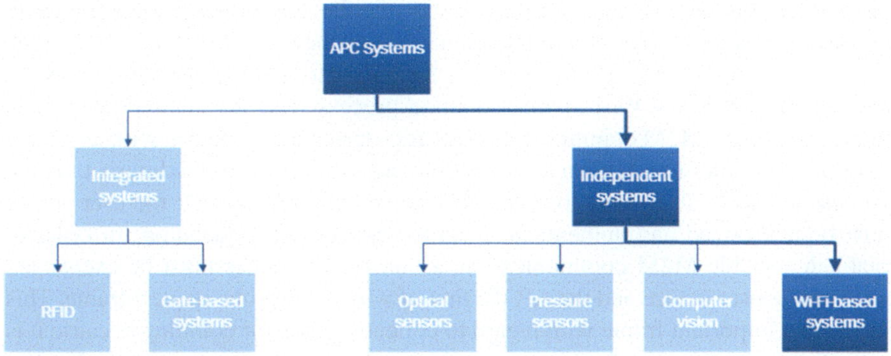

Fig. 12.1 Area and scope of research

the transport vehicles in the fleet. The counting is performed in "rotation," meaning that the vehicles with counters can be moved and swapped on other lines if needed. This implies that counting components are removed and implemented in individual PPT lines as needed, and it is estimated that approximately 10% of the complete vehicle fleet has APC components implemented in vehicles at any given time as described in report [18].

Paper [4] presents an overview of the most used APC solutions and it was noticed that, largely represented, counting methods are based on Wi-Fi technology. It was concluded that the reason for that is the widespread use of devices that support 802.11 standards (2018. 22 billion and predictions are that in 2025 the number will grow to 38.6 billion and in 2030 to 50 billion) and they are used to represent and determine the number of passengers in PPT vehicles [19]. Therefore, the identified area of interest is based on Wi-Fi technology and is shown in Fig. 12.1.

As mentioned, APC solutions based on Wi-Fi technology do not have universal application in PPT vehicles and one of the reasons that was determined in pilot experiments was the vast pool of data that can be gathered and all the different possibilities offered by such a solution. So, certain characteristics and collected data from 802.11 devices that are exploited to determine the number of passengers, i.e., devices in the vehicle, will be defined further in the section and the concept of operation for APC solutions based on Wi-Fi technology will be described.

3.1 WLAN Overview

A fundamental requirement for modern digital society, especially for concepts like Smart City and Society 5.0, is the reliability and availability of information any time and any place. Such requirements raise the need for certain standards of mobility, flexibility, ease of implementation, reasonable prices, security, and unobtrusiveness, which are a large part of the WLANs [20]. In accordance with that, the main goal of WLAN is to provide mobile and secure access to a network for mobile devices or users via wireless network media. With that said, counting mobile devices or users via Wi-Fi technology is based on the process of collecting probe request frames in the association of service communication between devices (Machine-to-Machine communication—M2M). This is why it is necessary to first explain this process as an essential part of the Wi-Fi-based counting process. The association service is a process in 802.11 M2M communication for registering a user device with an access point (AP) in order to establish a connection and gain access to the Internet via the wireless network. To be able to send and receive data over the AP, the user device must be authorized and authenticated via the process of association. In order to establish a viable M2M communication, a number of frames must be exchanged between the user device and the AP, the first of which is a probe request frame. This is the most important frame with regard to collecting data for counting because it is transmitted frequently and with no input from the user, but has some useful data stored in it (MAC address, timestamp, SSID, IBSS, etc.) [21].

The process of association can begin when the network is identified by the user device while scanning the surrounding environment. This mostly happens automatically, as was mentioned, because it transpires frequently and with no user input or interference. However, this process can be divided into active and passive modes, depending on the user device sending the probe request frames or waiting to receive beacon frames. So, active scanning is the process in which a user device transmits a probe request frame on every channel and then waits for the response from the nearby AP. The 802.11 standard has no fixed time defined for how long the user device has to wait for the response from the AP before sending a new batch of probe request frames. However, the usual time period is estimated to be 10 milliseconds. A simplification of the aforementioned process is shown in Fig. 12.2 [22].

Passive scanning, on the other hand, is a process in which the user device does not search for the nearby AP or initiate the communication, but listens instead for the transmitted beacon frames on available channels. These beacon frames are being transmitted by the AP every 100 milliseconds which makes it possible for the user device to miss them because it had not listened for long enough on the right channel. This is why active scanning can be considered better for the process of collecting data [22].

3.2 Concept for Collecting Data from PPT Using Wi-Fi

With everything described in the last section, the active scanning method is better for counting passengers in PPT and, as mentioned earlier, the probe request frames that are transmitted every 10 milliseconds are filled with potentially useful data. However, in the research for this paper, the authors have observed that some solutions use probe request frame collecting for their raw data gathering, but keep referring to it as passive scanning. The reason for this terminology difference is because some researchers and authors of papers [8–10, 13] consider their scanning to be unobtrusive for the passengers and so consider it passive. This conflict in terminology is probably caused by ambiguity since it is active from WLAN and network

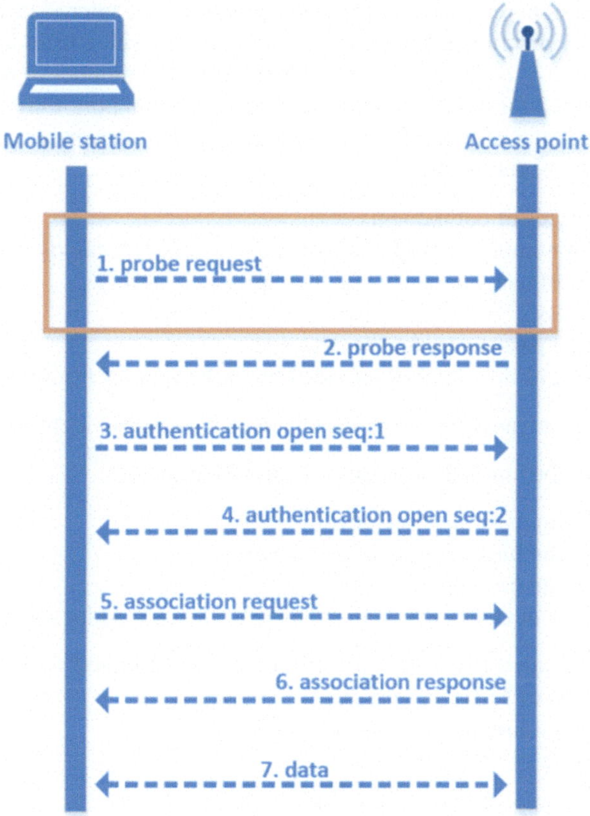

Fig. 12.2 Process of generating probe request frames and the focus for passenger counting

security perspective, but unobtrusive, noninvasive, and indirect, i.e., passive from the passenger counting perspective. For the purpose of this paper and counting passengers in PPT, the probe request frame collecting will be considered a passive method because of the unobtrusive nature toward the users.

Taking this into consideration, passenger counting can be performed at several levels depending on the purpose for which the data are collected. Therefore, it can be divided into

- System level: Making decisions on budgets and financial planning.
- Route or line level: Planning and scheduling the means of transport.
- Travel level: Determining whether to add or remove a specific PPT line;
- Stop level: Determining if and where to create a station.
- Segment level: Planning the transportation service [18].

In cases of smaller service areas or when it comes to a smaller PPT service provider, passenger counting is done at the "*macro*" level. This implies a systematic and route/line approach. However, the greater the complexity of the PPT infrastructure, the more passenger providers count at the "micro" levels (more precisely, previously mentioned travel, stop, and segment levels). In addition, the more detailed the "micro" levels are, the more demanding the number of passengers and the greater the need for technical and human resources support. Furthermore, based on examined existing research [6–17] some pros and cons were identified. Pros of Wi-Fi-based APC solution are

- Affordable prices of components.
- Ever-growing number of Wi-Fi capable devices in everyday use.
- Forgetfulness of users to deactivate their Wi-Fi capability.
- Unobtrusiveness.
- Does not make passengers feel monitored.
- Large amounts of information provided by probe request frame.

Despite stated advantages, some disadvantages were also identified:

- Difference in the number of passengers and user devices.
    - Passenger can have just one device.
    - Passenger can have more than one device.
    - Passenger can have no devices.
    - Detection of devices outside the vehicle or designated location.
- Need for calibration because of the difference in passengers and devices.
- Randomization of MAC addresses.
- Variations in sending probe request frames.
- Dynamic nature of PPT environment.
    - Variable distances of PPT stops and stations.
    - Variable waiting time of PPT vehicles.

Also, there are many different uses for data collected in the previously described ways in accordance with the mentioned research, ridership being the primary one, but with all the valuable data gathered, other uses can be extrapolated—OD matrices, waiting times, peak hours and behavioral trend, degree of congestion, etc. Most of that data can be formulated in some form of PmA results that help discover and analyze trends in passenger behavior. This, of course, greatly helps in designing, creating, running, optimizing, and maintaining PPT systems, infrastructure, networks, and vehicles.

3.3 Wi-Fi-Based APC

As already established, the most practical way to collect data with an APC based on Wi-Fi is to use active scanning because of the previously described probe request frame collection process that allows for flexibility when defining whether or not the user devices are inside the vehicle. For example, automatic probe request frame counting solution is practical in cases where the flow of people is dynamic, i.e., where people are passing by or staying just temporarily (public transport vehicles) because it has a relatively low probe request transmission time and can be manually adjusted if needed. It is also good for use in cases where the flow of people is static, i.e., where people stay in one place for a longer period of time (long-distance travel, PPT stations, etc.). In these circumstances, it is predicted that their devices will actively seek AP access.

Passive scanning, on the other hand, is quite impractical for counting people in general, because of its slow beacon frame transmission rate and channel distribution. This means that the counting error and reliability of the system suffer greatly and can be increased by the fact that the user device has moved out of range before it had the time to receive the beacon frame and send an answer.

According to all of these, it can be concluded that the greatest benefit when counting passengers in PPT vehicles is obtained by implementing an APC solution based on the probe request processing and analysis. The APC solution that uses probe request frames (active scanning) is currently the best usage of Wi-Fi technology since sending a frame is faster than sending a beacon and has enough information in the said frame to determine the number of user devices. The architecture of the system based on Wi-Fi consists of several components, but the counting concept is the same for most frames. The logical representation of the Wi-Fi-based APC can be seen in Fig. 12.3.

With everything already described, the user devices transmit probe request frames for the purpose of finding wireless networks that they can connect to. Generally, the components detect the frame and store it in an internal or external database. Once the frames are stored they must be filtered and calculated in a data management system. After processing, the results are stored (generally in some

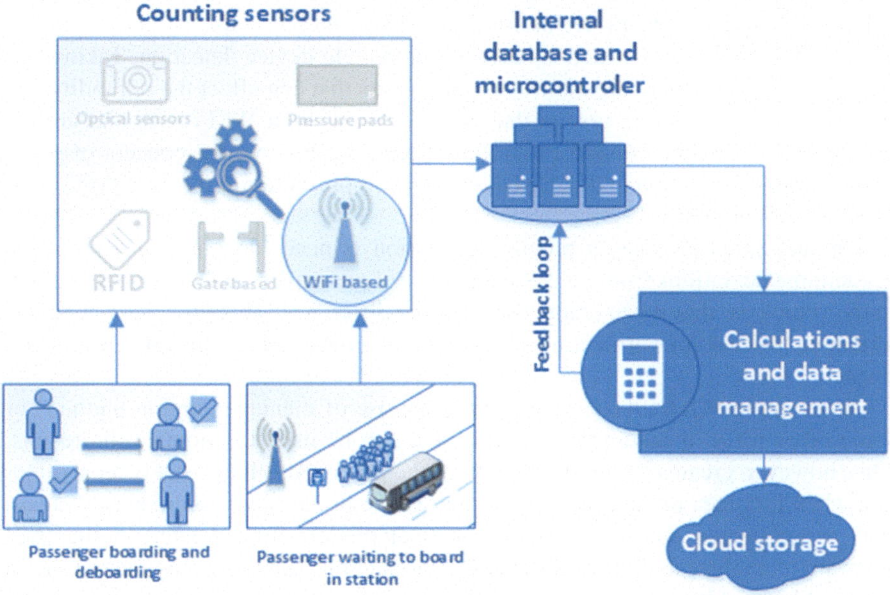

Fig. 12.3 Architecture of Wi-Fi-based APC solution

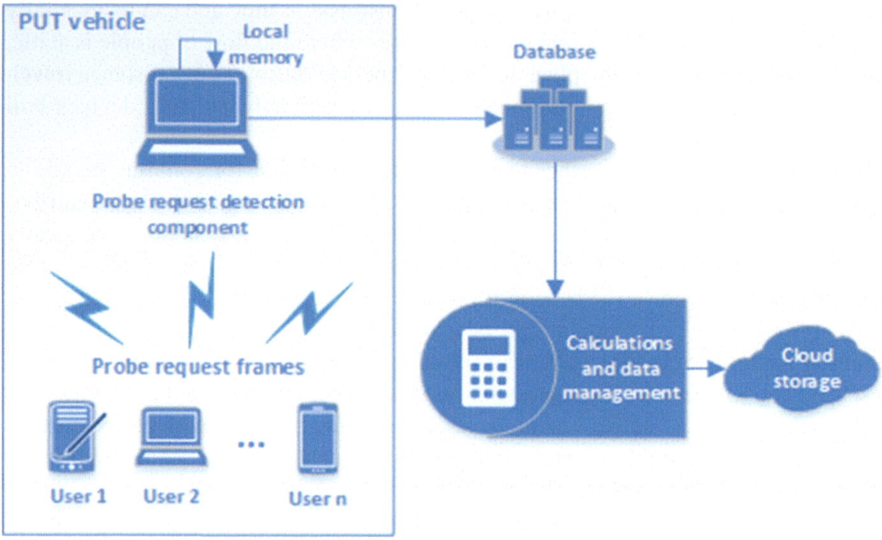

Fig. 12.4 Concept of probe request-based counting

cloud storage system) and sometimes used in a feedback loop to calibrate the components for the future measurements. The stored data should be ready and accessible when requested for analysis and decision-making process.

It is important to note that no APC solution or counting method provides 100% reliability or precision and the main disadvantage, as mentioned earlier, is the difference in the number of entries and exits to and from the vehicle. However, APC solutions based on Wi-Fi technology are distinctive because of their, already described, core concept which is based on in-vehicle device detection. Taking this into consideration, there are numerous parameters that can affect the reliability and precision of the counting results that are specific to their Wi-Fi base and are the reason for the aforementioned calibration. Figure 12.4 shows the concept of probe request frame collection and is based on logical representation for most APC solutions, as shown in Fig. 12.3.

For the stated reasons, a precise calibration is necessary and must take into account the deviations from the actual number of passengers and the signal strength in the vehicle so that it can compensate for the difference. This procedure connotes a reliable process for verification of the collected data and calculated or estimated results.

An example of a calibration process is the use of manual counting methods or some other form of automatic counting that can be used to compare the results. This, however, creates a hybrid variation of the proposed solution but is most effective when it comes to the precision of the results. Some other methods for "*cleaning*" the collected data are certain statistical models implemented in the used software to analyze and process the data (for example, counting how many times a unique MAC address is repeated, or defining a signal strength threshold to determine if the device/person is inside the designated counting zone).

4 Analysis of the Existing Research Results

Consequently to everything already stated in this research, Wi-Fi passenger counting uses the gathering and processing of data obtained from the probe request framework as the basis for counting user devices/passengers and the methods can differ because of the different conditions and possibilities in which the authors conducted their experiments. Therefore, this section presents an overview of the related literature and the research done on Wi-Fi-based APC and the analysis of the results of specific existing research from the field of APC solutions based on Wi-Fi. An overview of the studies related to the solution of Wi-Fi-based APC in public urban transport is shown in Table 12.1. Columns in the table represent what was considered important information of the researched papers.

The first column contains the name of the author, year of publication, and the referenced paper. The second column represents the mode of transport on which the particular solution was based (bus, train, or none if in laboratory conditions) and the third column, split into two, was created to show where the equipment was implemented (inside or outside the vehicle). The next two columns represent the results the authors estimated and/or calculated in their papers and the last column provides an overview of the used methods. Therefore, Table 12.1 demonstrates how many different variants are possible when using Wi-Fi as the basis of the APC solution.

As described in the previous sections, the core concepts of collecting probe request frames or measuring the signal strength can deliver datasets that can be used for any number of different possibilities—ridership, timestamps, OD matrices, waiting times, boarding and deboarding, peak times and values, length of stay, passenger grouping, mean absolute error, median absolute error, etc. So, alongside ridership data can be used for generating a lot of other information and any form of PmA to optimize, modernize, or enhance the already existing PPT systems or help in the creation of a new one.

Most research that was analyzed was conducted in buses, some were conducted in trains/metro, and some in laboratory conditions at universities or some other enclosed environment (office, large hall, etc.). It is not necessary to state that the experiments in dynamic real-life conditions, in PPT systems and vehicles in papers [6–9, 15, 23–29] are more relevant and have higher reliability status than those that were conducted in the controlled conditions with volunteers [10–14, 16, 17].

However, the aim of this research is to demonstrate that there are many different methods and analytical processes for collecting and interpreting the gathered data from the Wi-Fi network. This information might not be useful for some developed countries that already have some form of APC solution integrated in their PPT system, but it can serve as a great overview for less fortunate transport service providers to help them choose a method for counting passengers using their existing Wi-Fi network. Also, this can be of great help to other researchers, who are about to start experimenting in this field, to easily find some relevant literature on the subject and review the already explored methods and processes.

Table 12.1 Overview of studies related to the solution of Wi-Fi-based Automatic Passenger Counting in Public Urban Transport

Authors	Mode of transport	Measuring device location		Estimation and calculation	Counting performed	Methods
		Inside	Station			
Seunghan et al. (2020) [6]	Bus	–	+	OD matrices, waiting times	+	Probe request frame collecting, Wi-fi signal strength, and video recording analysis
Hidayat et al. (2020) [8]	Bus	+	–	OD matrices, mean absolute error, median absolute error, standard deviation of absolute error, root mean square, error, R2 score	+	Probe request frame collecting
Ribeiro et al. (2020) [9]	Bus	+	–	Number of passengers, peak values, and OD matrices	+	Probe request frame collecting
Nitti et al. (2020) [23]	Bus	+	–	Number of passengers, bus running time	+	Probe request frame collecting
Moser et al. (2019) [24]	Bus	+	–	Number of passengers	+	Probe request frame collecting
Oliveira et al. (2019) [16]	/	–	–	Mean relative error	+	Probe request frame collecting
Hakegard et al. (2018) [7]	Bus	+	–	OD matrices	+	Probe request frame collecting
Das et al. (2018) [11]	/	–	–	Passively encountering group (PEG)	+	Probe request frame collecting
Hidayat et al. (2018) [25]	Bus	+	–	Non-passenger data	+	Probe request frame collecting
Ribeiro et al. (2017) [26]	Bus	+	+	OD matrices, number of passengers	+	Probe request frame collecting
Song et al. (2017) [27]	Train	–	–	Waiting times, spectral clustering approach	–	Probe request frame collecting
Song et al. (2017) [13]	/	–	–	Length of stay (LOS), passenger flow, and occupancy	+	Probe request frame collecting
Hakegard et al. (2017) [28]	Bus	+	–	Number of passengers	+	Probe request frame collecting, signal strength analysis
Kang et al. (2016) [29]	Bus	+	–	OD matrices, mobile device tracking	+	Probe request frame collecting
Oransirikul et al. (2014) [15]	Bus	+	+	Number of passengers	+	MAC address collecting

5 Conclusion

Counting passengers using various technologies and methods, including Wi-Fi, is being increasingly introduced in the running and maintaining the public transport systems, networks, and vehicles. By analyzing the available literature and research, it was found that there is currently no clear taxonomy of selecting the optimal method or methodology for counting passengers using Wi-Fi technology. The main difference is in the approach and is shown depending on the location of the counting equipment, the type of vehicle and mode of transportation, the estimated results and actual measured values, the chosen method for filtering and analyzing the gathered raw data, etc. An important role in the APC solutions based on Wi-Fi technology has the understanding of the scanning methods and the difference between active and passive scanning and counting by different authors, which can be characterized as a potential challenge and a problem when it comes to terminology and taxonomy. This is why the authors of this paper made a decision to classify all the unobtrusive, noninvasive, and indirect methods as passive from the perspective of counting passengers in PPT.

In the future research, the authors expect to conduct their own pilot experiment for counting passengers, i.e., counting in a real dynamic traffic environment (PPT vehicle in motion, at a stop, at the terminal turn stop, in a garage, etc.) oriented to active and passive scanning using Wi-Fi technology. The planned experiment will be conducted in the city of Zagreb in Zagrebački Električni Tramvaj—ZET (Zagreb Electric Tramway)—transport network, and public transport vehicles.

References

1. Statista (2020) Passenger mileage in EU-28 in 2018, by mode (in billion passenger kilometers) [Internet]. https://www.statista.com/statistics/280519/passenger-mileage-in-eu-27-by-mode/. Last accessed 23 Dec 2020
2. Statista (2020) Share of passenger mileage in EU-28 in 2018, by mode [Internet]. https://www.statista.com/statistics/280520/share-of-passenger-mileage-in-eu-27-by-mode/. Last accessed 27 Dec 2020
3. Statista (2014) Percentage of passengers using transport in the EU in September 2014, by mode of transport [Internet]. https://www.statista.com/statistics/429068/modes-of-transport-used-by-eu-citizens/. Last accessed 28 Dec 2020
4. Grgurević I, Juršić K, Rajič I (2020) Review of automatic passenger counting Systems in Public Urban Transport. In: Proceedings of 5th EAI international conference on Management of Manufacturing Systems, EAI MMS 2020, cyberspace
5. Statista (2019) Number of smartphone users worldwide from 2016 to 2021(in billions) [Internet]. https://www.statista.com/statistics/330695/number-of-smartphone-users-worldwide/. Last accessed 29 Dec 2020
6. Seunghan R, Byungkyu BP, El-Tawab S (2020) WiFi sensing system for monitoring public transportation ridership: a case study. KSCE J Civ Eng 24:3092–3104
7. Håkegård J, Myrvoll TA, Skoglund T (2018) Statistical modelling for estimation of OD matrices for public transport using Wi-fi and APC data. In: 2018 21st international conference on intelligent transportation systems (ITSC), pp 1005–1010
8. Hidayat A, Terabe S, Yaginuma H (2020) Estimating bus passenger volume based on a Wi-fi scanner survey. Transportation Research Interdisciplinary Perspectives 6:100142

9. Ribeiro M, Galvão B, Prandi C, Nunes N (2020) Passive Wi-fi monitoring in public transport: a case study in the Madeira Island. In: Proceedings of TRA2020, the 8th transport research arena: rethinking transport towards clean and inclusive mobility, Helsinki, Finland
10. Ooi Y, Wai KZ, Tan I, Sheng OB (2016) Measuring the accuracy of crowd counting using Wi-fi probe-request-frame counting technique. J Telecommun Electron Comput Eng 8:79–81
11. Das S, Chatterjee S, Chakraborty S, Mitra B (2018) An Unsupervised Model for Detecting Passively Encountering Groups from WiFi Signals, 2018 IEEE Global Communications Conference (GLOBECOM), Abu Dhabi, United Arab Emirates, pp. 1–7
12. Yoshida, T., Yoshiaki, T.: Estimating the number of people using existing WiFi access point in indoor environment. Adv Comput Sci, pp. 46–53 (2015)
13. Le TV, Song B, Wynter L (2017) Real-time prediction of length of stay using passive Wi-fi sensing. In: 2017 IEEE international conference on communications (ICC), Paris, pp 1–6
14. Brandon S (2015) Estimating passenger flow & occupancy on board public transport buses through mobile participatory and opportunistic sensing. University of Dublin, Trinity College, Dissertation
15. Oransirikul T, Nishide R, Piumarta I, Takada H (2014) Measuring bus passenger load by monitoring Wi-fi transmissions from Mobile devices. Proc Technol 18:120–125
16. Oliveira L, Schneider D, De Souza J, Shen W (2019) Mobile device detection through WiFi probe request analysis. IEEE Access 7:98579–98588
17. Oliveira L, Henrique J, Schneider D, de Souza J, Rodriques S, Sherr W (2018) Sherlock: capturing probe requests for automatic presence detection. In: IEEE 22nd international conference on computer supported cooperative work in design (CSCWD), Nanjing, pp 848–853
18. Boyle DK (1998) Passenger counting technologies and procedures. Transit cooperative research program (TCRP) synthesis 29, published by. Transportation Research Board, Washington
19. Statista (2019) Number of internet of things (IoT) connected devices worldwide in 2018, 2025 and 2030 (in billions) [Internet]. https://www.statista.com/statistics/802690/worldwide-connected-devices-by-access-technology/ last accessed 2020/12/30
20. Gast MS (2005) 802.11 wireless networks: the definitive guide, 2nd edn. O'Reilly Media, Sebastopol, p 672
21. Cisco Press Homepage, https://www.ciscopress.com/articles/article.asp?p=1271797&seqNum=2 last accessed 2020/12/30
22. Freudiger J (2015): How talkative is your mobile device?: an experimental study of Wi-Fi probe requests. WiSec '15: Proceedings of the 8th ACM Conference on Security & Privacy in Wireless and Mobile Networks, WiSec'15: 8th ACM Conference on Security & Privacy in Wireless and Mobile Networks New York New York, pp 1–6
23. Nitti M, Pinna F, Pintor L, Pilloni V, Barabino B (2020) iABACUS: a Wi-fi based automatic bus passenger counting system. Energies 13(6):1–21
24. Mehmood U, Moser I, Jayaraman PP, Banerjee A (2019) Occupancy estimation using Wi-fi: a case study for counting passengers on busses. In: IEEE 5th world forum on internet of things, pp 165–170
25. Hidayat A, Terabe S, Yaginuma H (2018) Determine non-passenger data from WiFi scanner data (MAC address), a case study: Romango bus, Obuse, Nagano prefecture, Japan. Int Community Spatial Planning Sustain Dev 6(3):154–167
26. Prandi C, Nunes N, Ribeiro M, Nisi V (2017) Enhancing sustainable mobility awareness by exploiting multi-sourced data: the case study of the Madeira Islands. In: Sustainable internet and ICT for sustainability (SustainIT), Funchal, pp 1–5
27. Song B, Wynter L (2017) Real-time public transport service-level monitoring using passive WiFi: a spectral clustering approach for train timetable estimation. CoRR abs/1703.00759
28. Myrvoll TA, Håkegård JE, Matsui T, Septier F (2017) Counting public transport passenger using WiFi signatures of mobile devices. In: IEEE 20th international conference on intelligent transportation systems (ITSC), Yokohama, Japan, pp 1–6
29. Kang L, Qi B, Banerjee S (2016) A wireless-based approach for transit analytics. In: HotMobile '16: proceedings of the 17th international workshop on Mobile computing systems and applications, pp 75–80

Chapter 13
Impact of Dynamic Characteristics of Gears on the Reliability of Prilling Equipment

Maryna Demianenko, Oleksandr Starynskyi, Ivan Pavlenko, Vsevolod Sklabinskyi, Justyna Trojanowska, Maksym Skydanenko, Oleksandr Liaposhchenko, and Vitalii Ivanov

1 Introduction

The creation of new granular materials in the active hydrodynamic environment is realized using highly efficient prilling equipment. Mainly, vibrational prillers are widely used in various industries [1]. They allow getting granular material, which has a high specific surface area, the possibility of long-term storage, dosage, and the like [2].

Recently, the studies of vibrating granulators and prillers aimed at determining the influence of oscillation frequencies on the workflow's efficiency are quite widespread. In this regard, vibrating granulators were developed to produce nitrogen fertilizers at the Department of Chemical Engineering (Fig. 13.1).

As a result of the studies, the influence of vibration characteristics and fluid flow through the openings on the change in pressure in the melt jet was determined, which allows predicting the jets' decay parameters to obtain monodisperse droplets, resulting in improved quality of the finished product. Particularly, a similar dealer was considered, and studies have made it possible to establish a relationship between the location of the source of forced oscillations in the inner space of the vibrating granulator by the diameter of the droplets and satellite droplets in the process of decay of liquid jets, and as a consequence, to improve the quality of the product obtained and to minimize the loss of the target product. Additionally, the problem of contamination of the perforated holes of the pellets was considered. It was proposed

M. Demianenko · O. Starynskyi · I. Pavlenko · V. Sklabinskyi · M. Skydanenko
O. Liaposhchenko · V. Ivanov (✉)
Sumy State University, Sumy, Ukraine
e-mail: ivanov@tmvi.sumdu.edu.ua

J. Trojanowska
Poznan University of Technology, Poznań, Poland

L. Knapcikova et al. (eds.), *Sustainable Management of Manufacturing Systems in Industry 4.0*, EAI/Springer Innovations in Communication and Computing,
https://doi.org/10.1007/978-3-030-90462-3_13

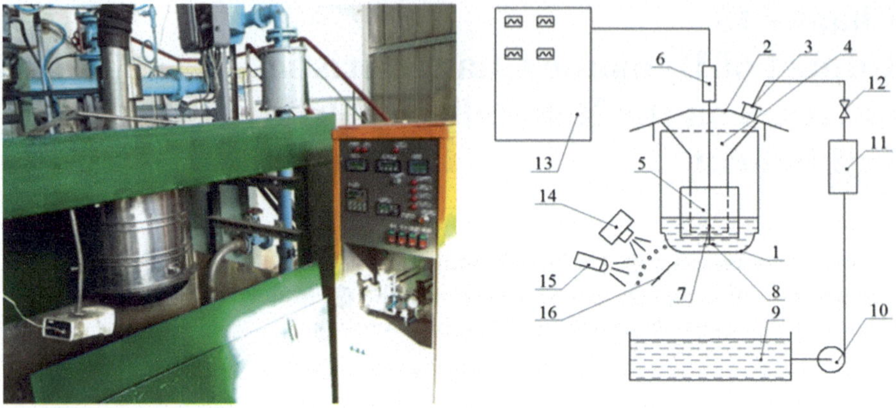

a b

Fig. 13.1 Appearance (**a**) and scheme (**b**) of the experimental stand: 1 -- basket; 2 – the case; 3 – branch pipe; 4 – collector; 5 – filter element; 6 – electromagnetic oscillator (actuator); 7 – stock; 8 – disk; 9 – buffer capacity; 10 – pump; 11 – flowmeter; 12 – the valve; 13 – control cabinet; 14 – photofixation device; 15 – strobe; 16 – scale screen

to impose high-frequency oscillations on the perforated bottom of the priller. Therefore, the vibration effects on the environment and the vibrating pelletizer's constituent elements are the most important criteria for ensuring their reliable and efficient operation.

It should be noted that the vibrational flow to the environment also depends on the smooth transmission of movement from the motor to the actuator. In the equipment produced by the world enterprises for the production of prilling equipment (e.g., "Kreber," "KCVT Ltd.," etc.), the dispersion is carried out both by the interaction of the jet of liquid with the gas stream and by the vibrations transmitted from the actuator to the working environment and the basket. For this purpose, particularly, tooth gears are used. However, while gearing is in operation, beating may occur due to the entry of each gear tooth and wheel into gear, adversely affecting the robot actuator and the pelletizer as a whole. Such a contribution to the vibration effects on the environment can significantly impair the vibrating granulation process efficiency and the target product's quality. Simultaneously, the transmission of vibration-driven gear is not a completely solved problem because of its complexity and interdisciplinary nature. Thus, this aims to investigate the possibilities of eliminating the problem mentioned above by ensuring the smooth operation of the gear pair and minimizing the beating.

2 Literature Review

Equipment efficiency depends on many parameters, including technological capabilities [3–6], materials and coatings [7, 8], energy efficiency indicators [9, 10], and quality [11, 12].

To ensure a smooth engagement and minimize the beating, tapered gears with a circular tooth line are used. One type of conical gears with a circular tooth line are gears with double-convex and double-concave teeth, with the gear wheel usually made with double-convex teeth and the wheel with double-concave teeth. The main advantage of those is reducing contact loads and the increase of smooth-running because they have different curvatures in two directions. Samoilov A. and Onyshchenko V. proposed such gear designs. The related research works were carried out by Srebnaya E., Kuznetsova A., and Klymenko V.

The analysis of gearing parameters can be carried out using software systems that implement the finite element method. Thus, in [13], gears' modeling using the finite element method is carried out. The contact loads on a standard and modified tooth model along the entire contact surface are determined. The influence of the lateral tooth surface shape on the rate of change of contact loads and the wheel's rotation speed is determined.

[14] presents a study of gears, which was carried out with the "ANSYS Workbench" software package. The purpose of the study was to determine the maximum and minimum contact loads on the surfaces of the teeth for three materials – cast iron, medium-carbon steel, and high-carbon steel, as well as for three forms of holes to reduce the load, circular, elliptical, and "aerofin" holes. In a study [15], a cylindrical tooth-to-gear transmission was modeled using Finite Element Modeling (FEM) in the "ANSYS" software package. As a result of the studies, the tooth's lateral surface contact loads were determined, and theoretical calculations confirmed the simulation results. [16] presented a simulation of a cylindrical oblique transmission using the "ANSYS Workbench" and Finite Element Analysis (FEA) software. They concluded that the helical gear pair's increased efficiency was attained by enhancing power transmission capacity and life would increase by calculating bend and contact stress and strength.

In [17], an analysis of a pair of cylindrical skew wheels was carried out using the FEM tool "Altair Opti Struct" solver. In the simulation of this gear, loads for stainless steel and Nylon 66 were determined. In [18], simulations of cylindrical slanting gear using FEA in the ANSYS software complex were presented. Compared to the AGMA procedure, the results obtained from the "ANSYS" software show that there is a slight deviation of 4.3%. It has also been argued that the prediction of bending load values at any required module and end width by ANSYS is reliable for gears. The maximum equivalent load of the tooth and the tooth's deformation were determined due to the simulations [19]. In a study [20], computer simulations of a cylindrical gear wheel in the ANSYS Workbench 14.0 software complex were performed, which resulted in the determination of bending loads for three different torques, and deformations of the tooth head were determined.

There are numerous works dedicated to profiling and the improvement of gears and gearboxes. In particular, the study [21] is devoted to the fault diagnosis of machines based on the Dempster–Shafer evidence theory. In [22], the artificial neural network data fusion classification technique was used for induction motor fault detection. The mathematical model of particle movement on a helical surface was developed in the research paper [23]. Article [24] is devoted to ensuring the reliability of steel parts' surfaces based on the alloying. The corresponding mathematical model for strengthening surfaces using the electroerosive alloying method was proposed in the research [25].

Additionally, in the article [26], the technique for ensuring the profile gear grinding modes was developed to provide uniform residual temperatures. Finally, the scientific works in ensuring the reliability of technological equipment in the related areas of the research are presented below. Remarkably, the article [27] presents a comprehensive analysis of the equipment with the combined heat and mass exchange. Papers [28, 29] present up-to-date trends in jet technologies with improved performance.

Finally, in the articles [30, 31], authors studied modified gear drives for multi-operational machines with increased load capacity and the toothed belt transmission with arched teeth with the corresponding application in the technological equipment [32].

3 Research Methodology

3.1 A Brief Overview of Bevel Gears

Bevel gears are widely used for the smooth transmission of rotational motion in industrial equipment and machinery. They can be classified by tooth shape as straight, circular, and tangential.

Bevel gears are more complicated to manufacture than cylindrical ones since it is necessary to maintain tolerances on the axle angles and tooth thickness. They are also more difficult to install due to the need to fix the axial position accurately and ensure the cones' vertices coincide. Simultaneously, there are standard methods of their design and manufacture-based specialized gear cutting tools and grinding equipment. Notably, its use is justified only in the case of serial production. Therefore, it is cost-effective to manufacture gears in a single production using multifunctional Computer Numerical Control (CNC) milling machine tools.

Remarkably, bevel gears have the following advantages: smooth and silent traction than other types of gears and minimal beating due to smooth grip.

3.2 Designing of Traditional Bevel Gears

Creating a 3D model of a biconvex gear and a biconcave wheel on CNC machine centers is associated with many difficulties. The "KOMPAS 3D" software package and its module "Shafts and mechanical transmissions" allow carrying out geometrical calculations. As a result, 3D models of typical bevel gears can be automatically built. Additionally, contact surfaces can be localized. All these allow achieving the required accuracy of the manufacture of the gears on CNC machine tools.

"KissSOFT" is software for designing, analyzing, and optimizing machine parts, e.g., gears, shafts, bearings, connecting elements, springs, belts, and chain transmissions. The program allows to accurately calculate the strength, reliability, and service life. The program has a broader choice of gears and the ability to calculate and automatically build 3D models accurately. It is also possible to modify the surfaces of the teeth.

It is possible to carry out geometrical calculations and generate 3D models of gears using the following software packages: "Autodesk Inventor," "SolidWorks," "GearTrax AI," "T-Flex CAD," etc. However, the abovementioned software does not currently offer reliable methods and algorithms for calculating geometric parameters and creating 3D models of bevel gears with biconvex and biconcave teeth. Therefore, 3D models should be built manually. This fact reduces the probability of obtaining accurate geometric dimensions of the transmission and contact surfaces. As a result, after creating a 3D gear model, it is mandatory to calculate the teeth' stiffness. Additionally, to study gear transmission dynamics, it is necessary to assess the tooth shape's influence on the load area's contact stiffness.

Given the above, it was suggested to check the smoothness, the magnitude of the beating, determine the contact surface and contact loads, and study the rigidity of the created 3D models of gear pairs, the sizes calculated by standard methods, and their improvement before the manufacture.

Based on the literature review, it can be argued that the use of the FEA method is appropriate for the study of gears, determination of contact loads, bending loads, stiffness, and deformation. Therefore, "ANSYS Workbench" software and its module "LS-DYNA," which is based on a finite element method and is commonly used for transient processes, was selected for the analysis of gearing parameters.

Increasing the smoothness and reducing the beating is possible due to the gear's high-precision design and manufacture. To achieve a high degree of accuracy of the dimensions controlled and measured, namely the tooth's thickness in the measuring section and the height up to the chord in the measuring section, gears, and wheels of such transmission, they are made on milling CNC machines. Creating a 3D model of a double-convex gear and double-concave wheel for CNC machine tools also has several difficulties. Particularly, CAD software like "Autodesk Inventor" and the module "Shafts and Mechanical Gears" of "KOMPAS-3D" allow performing geometric calculations, which can automatically build 3D models of typical conical gears, as well as localize the contact surfaces, which will achieve the required precision in the manufacture wheels and gears on CNC machines.

The "KissSOFT" software offers a broader range of gear types, the ability to accurately calculate and automatically build 3D models, as well as the ability to modify gear tooth surfaces. However, the above programs do not currently offer reliable methods or algorithms for calculating geometric parameters and creating 3D conical models with double-convex and double-concave teeth, so 3D models should be built manually. It reduces the likelihood of obtaining accurate geometric dimensions of the transmission and the contact surfaces of the lateral surfaces of the teeth and a workable finished gear. Therefore, after creating a 3D model of such gear, it is necessary to calculate the teeth' rigidity. To investigate the gearing dynamics, it is essential to evaluate the effect of the shape of the tooth profile on the contact stiffness in the load zone.

3.3 *Designing of Specialized Bevel Gears*

Based on the capabilities of up-to-date CAD/CAE systems (i.e., to create 3D models and software that implements the finite element analysis), the following methodology of designing a bevel gear-pair with biconvex and biconcave teeth is proposed.

At the first stage, carrying out geometrical calculation of a bevel gear-pair by standard methods is offered. In this case, the output contour of the tooth for bevel gears is performed by GOST 16202-81 with the circular teeth. The output contour is chosen since the biconvex and biconcave teeth profiles are also built based on it. According to the geometric calculation results, 3D models of wheels and gears with circular teeth are generated. After, gear and a wheel's teeth in three sections are generated considering the inner and outer ends and measuring area.

At the second stage, 3D models of a biconcave wheel and a biconvex gear are built from the obtained tooth profiles. They are also fixed in space. The vertices of the cones must coincide since the wheel and gear should be engaged, which is essential for further computer simulations.

At the third stage, numerical simulation is performed using the finite element method of a gear-pair according to the following algorithm:

1. Design the 3D calculation grid.
2. Setup initial and boundary conditions (i.e., selection of contact surfaces and axial velocity).
3. Make calculations and analyze the results.

Finally, the conclusion about gear-pair reliability, size of contact loadings, beating, and a tooth pair's contact surface is made. In the case of unsatisfactory reliability indicators, it is needed to return to the first stage, make changes to the original data and repeat all the stages until satisfactory results are obtained. Figure 13.2 presents the proposed methodology.

4 Results

4.1 *Development of 3D Models of Gears and Wheels*

To create 3D models of bevel gear with a circular tooth, a software package "KOMPAS 3D" was chosen, particularly its module "Shafts and Mechanical Transmissions." Firstly, a geometric calculation of bevel gear with a circular tooth was performed. The initial contour was selected based on the standard "GOST

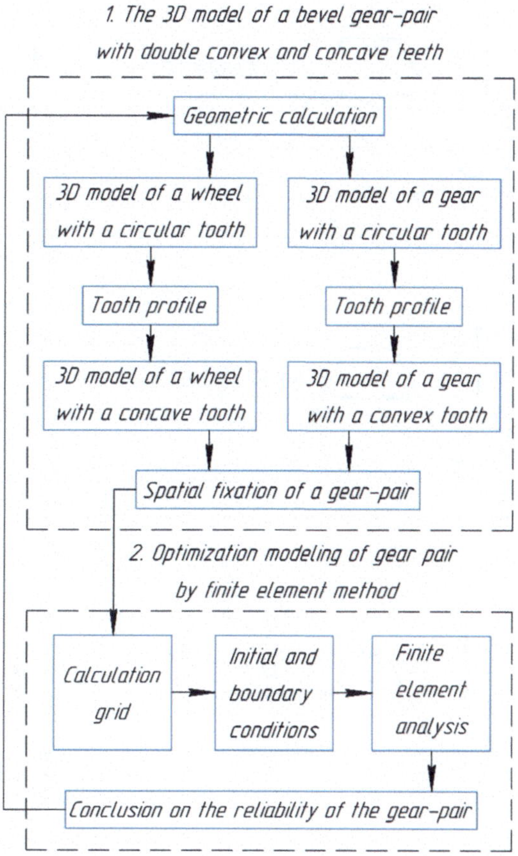

Fig. 13.2 The proposed methodology of designing bevel gears

16202-81." Also, the tooth's axial shape was selected, in which the teeth are proportionally reduced, and the top of the dividing cones have coincided. In this case, the tooth height is proportional to the conical distance.

Notably, since the curvature radius has the side surfaces of biconcave and biconvex teeth, a specialized diameter of the gear cutting tool, i.e., 457.2 mm was set.

As a result of the calculation, the bevel wheel's geometric parameters and gear with circular teeth were obtained. After that, solid models of wheels and gears were generated. The teeth' profiles for the wheel and gear were obtained in three sections from these solid models. These models will be used for the formation of the teeth lateral surfaces (Fig. 13.3).

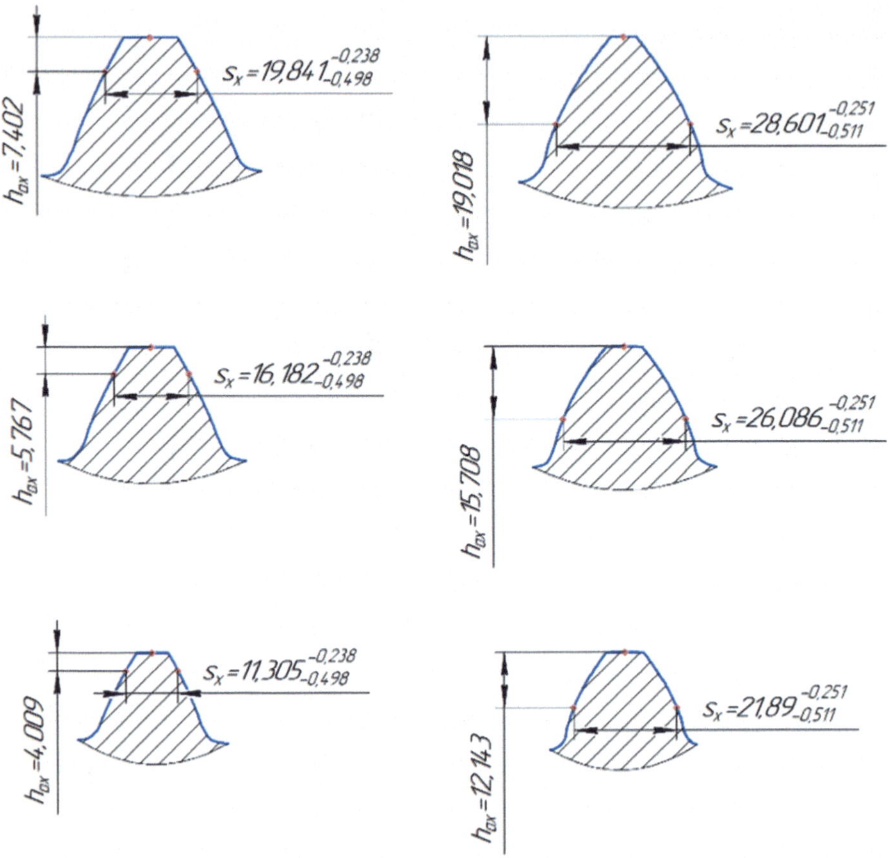

Fig. 13.3 Profiles of circular teeth of a conical gear-pair

4.2 *Designing of the 3D Model of a Biconvex Gear and a Wheel*

The creation of a 3D model of the bevel wheel and gear was carried out in several steps. Initially, a cylindrical stage was built, and a conical stage will form a toothed crown. Then three initial contours of a tooth were built based on circular teeth profiles, particularly in the measuring section.

Next, these contours were combined considering the gear head's geometry, which forms the lateral surfaces of the teeth. Based on the operation "Cutting," a conical cavity between the left- and right-side surfaces of two adjacent teeth was built. Finally, the gear crowns of a wheel and gear were built using the previously formed array. The results of 3D modeling are presented in Fig. 13.4.

For further computer simulation, the above 3D models were fixed in space. In this case, local coordinate systems were created at the vertices of the dividing cones.

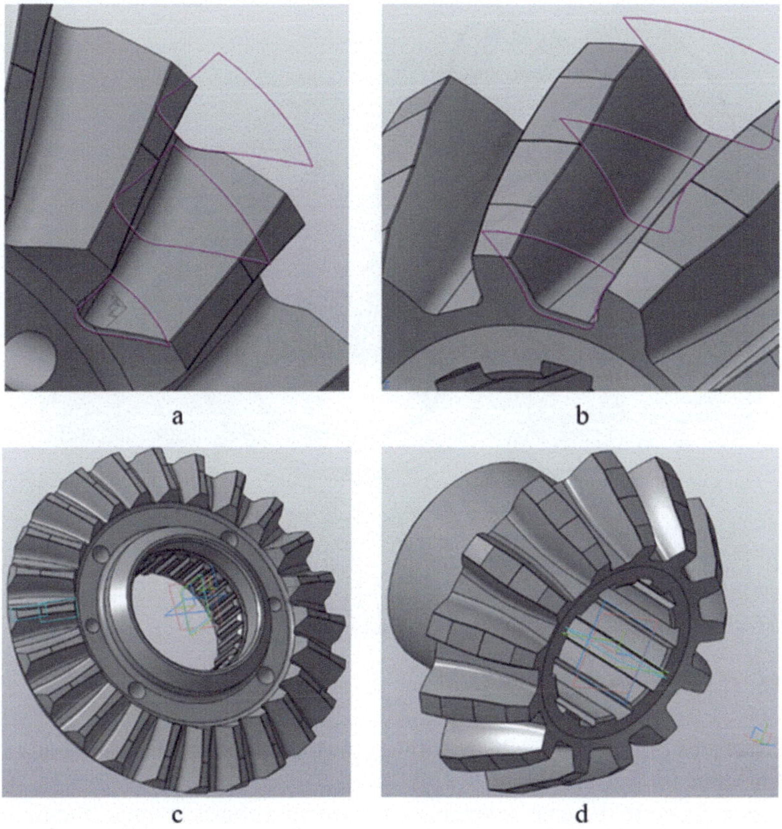

Fig. 13.4 Output contours of teeth (**a**, **b**) in three sections, and the generated 3D models (**c**, **d**): **a**, **c** – bevel wheel; **b**, **d** – bevel gear

The lase ones are fixed using the operation "Coincidence" and at an angle relative to each other. Consequently, the teeth are engaged (Fig. 13.5).

In connection with those mentioned above, ANSYS Workbench software, namely its Explicit Dynamics module, was used to investigate the double-convex and double-concave tooth geometry of the tooth gears. Of course, the first step was to build a 3D computational geometry. As noted above, its automatic generation using COMPASS 3D and KissSOFT is not possible. That is why we will perform construction in several steps.

In the first step, we construct three initial contours of the tooth according to the standard "Basic requirements for interchangeability. Conical gearings with circular gears. Basic rack," namely in a measured section, considering for gear, only the convex part of it, symmetrically reflecting it relative to the axis; for a wheel, only its curved part, symmetrically reflecting concerning the axis. The second step aims to combine the contour data while considering the milling geometry by which the

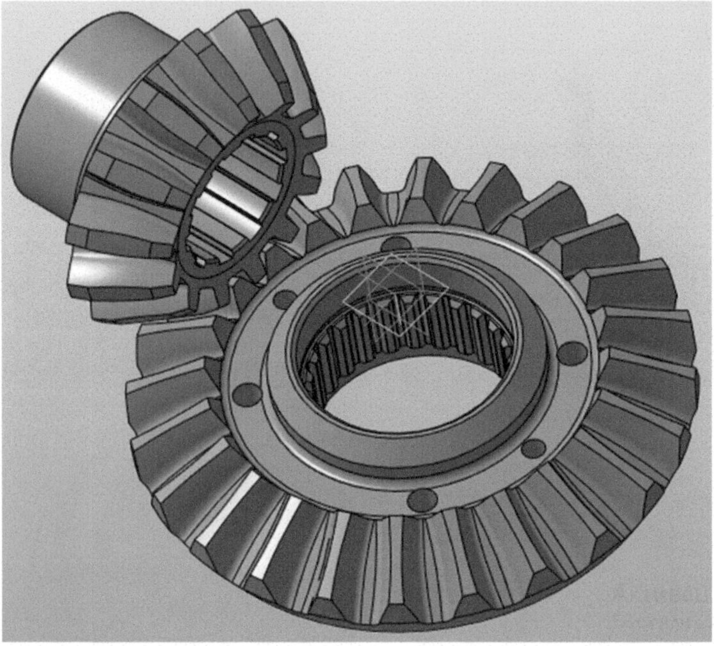

Fig. 13.5 The 3D model of the bevel gear-pair

lateral surfaces of the gear teeth are formed. At the final step, the main body of the wheel is built.

The next stage in the gear's dynamics study is the construction of the finite element calculation grid. Particular attention should be paid to the contact area of the gear wheels. The Contact Sizing tool was used for this purpose. The maximum element size was set to 1 mm in the contact area. In the general settings, the method of constructing a finite element curvature grid with an average element size of 5 mm was chosen.

4.3 *Initial and Boundary Conditions and the Finite-Element Grid*

The next stage in studying the dynamics of gears is designing a finite element model. At this stage, particular attention should be paid to the setup of contact interactions. The "Contact Sizing" tool was used for this purpose. In the contact area, the maximum element size is equal to 1 mm. In the general settings, the method "Curvature" for designing a finite element calculation grid was chosen. The average

Fig. 13.6 3D finite-element grid for the bevel gear-pair

size of the element is 5 mm. The obtained 3D finite-element grid is shown in Fig. 13.6.

The next step is setting up a contact. The Frictional type was chosen; the friction coefficient was set to 0.15; the wording – Augmented Lagrange. Body interaction (Type – Frictional) was also configured. The boundary conditions were chosen to secure the wheel hub on all degrees of freedom except rotating about the axis. Gear speed of 250 Hz corresponds to the range of the actuator of the granulator.

One of the calculation's essential steps is the Analysis Setting configuration: Maximum Numbers of Cycles: $1 \cdot 10^7$; End Time – 28 ms (one full turn time); Solver Type – Low Velocity.

4.4 Numerical Simulation Results

As a result of the simulation in the LS-DYNA module of the ANSYS software, the iso-surface displacements shown in Fig. 13.7 are obtained.

Additionally, it is advisable to analyze the contact surface area and the averaged force over the contact surface over time to check the gear pair engagement. This dependence is shown in Fig. 13.8.

In this case, the clash is not observed, and the process of interaction of the gear wheels can be divided into two stages. In the first stage, the steady-state mode is

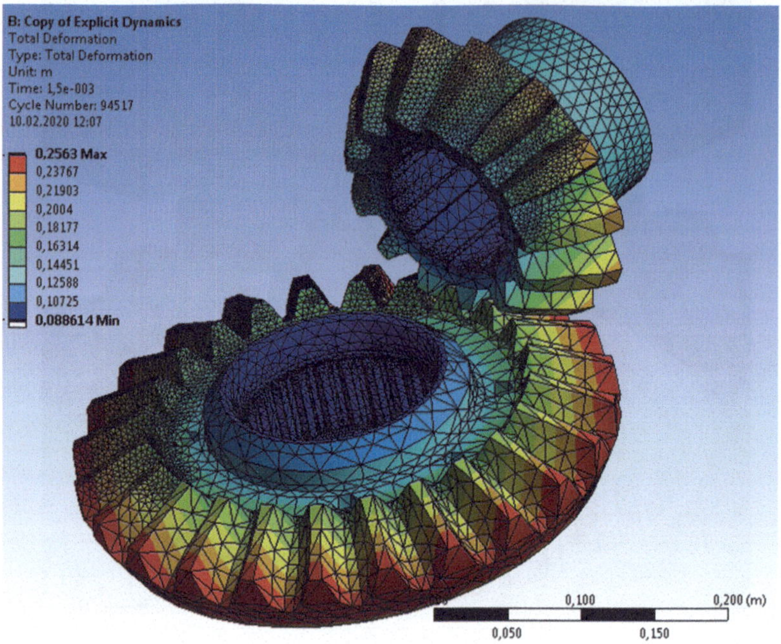

Fig. 13.7 The explicit dynamics of the gear-pair

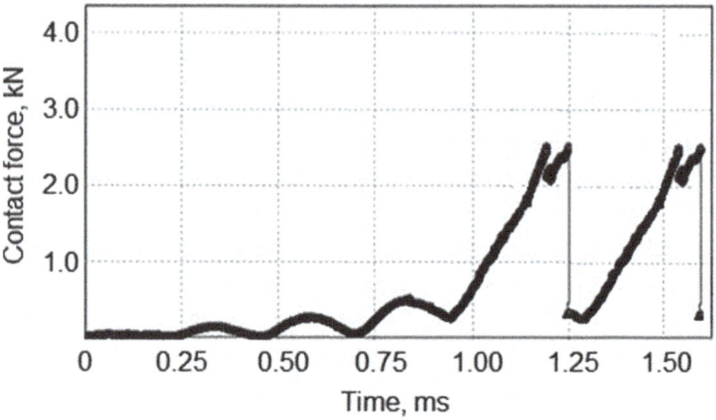

Fig. 13.8 The nonstationary contact force

reached, in which the force of contact interaction has a smaller amplitude compared to the next stage. During the second stage, the normal operation of the gear pair occurs.

As can be seen from Fig. 13.8, when the gear pair enters the operating mode, the amplitude of the oscillation of the contact force increases, with a further sharp decrease. It results in a slight jump in contact force compared to the basic amplitude.

5 Conclusions

Thus, the article shows that the nonstationary contact force that arises in the process of a clutch of the gear teeth has a periodic, saw-like form. This result indicates the presence of a polyharmonic response, the fundamental harmonic of which has a frequency equal to the product of the speed of the gear and the number of teeth. Moreover, along with the fundamental harmonic, there are pronounced superharmonics in the frequency spectrum. These components of the contact force adversely affect the monodispersity of the granules obtained in the vibrating granulator.

Thereby, the presence of clear polyharmonic response, the inability to adjust the process flexibly, makes it inexpedient to use gears as actuators of prilling equipment. However, more flexible V-belt transmissions are devoid of the disadvantages above. They can be used as reliable actuators that have proven themselves well in the granulation equipment market and have been successfully implemented at the Department of Chemical Engineering of Sumy State University.

Acknowledgments This research was partially funded by the Research and Educational Center for Industrial Engineering (Sumy State University), International Association for Technological Development and Innovations, and Research Grant "Creation of new granular materials for nuclear fuel and catalysts in the active hydrodynamic environment" (State reg. no. 0120U102036) ordered by the Ministry of Education and Science of Ukraine. The part of the results has also been obtained within the project "Fulfillment of tasks of the perspective plan of development of a scientific direction "Technical sciences" Sumy State University" funded by the Ministry of Education and Science of Ukraine (State reg. no. 0121U112684).

References

1. Skydanenko M, Sklabinskyi V, Saleh S, Barghi S (2017) Reduction of dust emission by monodisperse system technology for ammonium nitrate manufacturing. PRO 5(3):37. https://doi.org/10.3390/pr5030037
2. Ostroha R, Yukhymenko M, Lytvynenko A, Bocko J, Pavlenko I (2019) Granulation process of the organic suspension: fluidized bed temperature influence on the kinetics of the granule

formation. In: Ivanov V et al (eds) Advances in design, simulation and manufacturing. DSMIE 2018, Lecture Notes in Mechanical Engineering. Springer, Cham, pp 463–471. https://doi.org/10.1007/978-3-319-93587-4_48
3. Voloshina A, Panchenko A, Boltyansky O, Titova O (2020) Improvement of manufacture workability for distribution Systems of Planetary Hydraulic Machines. In: Ivanov V et al (eds) Advances in design, simulation and manufacturing II. DSMIE 2019, Lecture Notes in Mechanical Engineering. Springer, Cham, pp 732–741. https://doi.org/10.1007/978-3-030-22365-6_73
4. Voloshina A, Panchenko A, Panchenko I, Titova O, Zasiadko A (2019) Improving the output characteristics of planetary hydraulic machines. IOP Conference Series: Materials Science and Engineering 708(1):012038. https://doi.org/10.1088/1757-899X/708/1/012038
5. Fesenko A, Yevsiukova F, Basova Y, Ivanova M, Ivanov V (2018) Prospects of using hydrodynamic cavitation for enhancement of efficiency of fluid working medium preparation technologies. Periodica Polytechnica Mechanical Engineering 62(4):269–276. https://doi.org/10.3311/PPme.11877
6. Fesenko A, Basova Y, Ivanov V et al (2019) Increasing of equipment efficiency by intensification of technological processes. Periodica Polytechnica Mechanical Engineering 63(1):67–73. https://doi.org/10.3311/PPme.13198
7. Kostyuk G (2019) Prediction of the microhardness characteristics, the removable material volume for the durability period, cutting tools durability and processing productivity depending on the grain size of the coating or cutting tool base material. In: Lecture notes in mechanical engineering. Pleiades Publishing, pp 300–316
8. Kostyuk G, Nechyporuk M, Kostyk K (2019) Determination of technological parameters for obtaining nanostructures under pulse laser radiation on steel of drone engine parts. In: 10th international conference on dependable systems, services and technologies, DESSERT 2019. Institute of Electrical and Electronics Engineers Inc
9. Kiyko S, Druzhinin E, Prokhorov O, Ivanov V, Haidabrus B, Grabis J (2020) Logistics control of the resources flow in energy-saving projects: case study for metallurgical industry. Acta logistica 7(1):49–60. https://doi.org/10.22306/al.v7i1.159
10. Kotliar A, Basova Y, Ivanov V, Murzabulatova O, Vasyltsova S, Litvynenko M, Zinchenko O (2020) Ensuring the economic efficiency of enterprises by multi-criteria selection of the optimal manufacturing process. Management and Production Engineering Review 11(1):52–61. https://doi.org/10.24425/mper.2020.132943
11. Lebedev V, Tonkonogyi V, Yakimov A, Bovnegra L, Klymenko N (2019) Provision of the quality of manufacturing gear wheels in energy engineering. In: Ivanov V et al (eds) Advances in design, simulation and manufacturing. DSMIE 2018, Lecture notes in mechanical engineering. Springer, Cham, pp 89–96. https://doi.org/10.1007/978-3-319-93587-4_10
12. Denysenko Y, Dynnyk O, Yashyna T, Malovana N, Zaloga V Implementation of CALS-technologies in quality management of product life cycle processes. In: Ivanov V et al (eds) Advances in design, simulation and manufacturing. DSMIE 2018. Lecture Notes in Mechanical Engineering, pp 3, 2019–12. https://doi.org/10.1007/978-3-319-93587-4_1
13. Markovic K, Franulovic M (2011) Contact stresses in gear teeth due to tip relief profile modification. Engineering Review 31(1):19–26
14. Potghan D, Sharma S (2015) Stress reduction by introducing stress relieving features of spur gear used in lathe headstock. International Journal of Engineering Sciences and Research Technology 8:182–189
15. Farhan M, Karuppanan S, Patil SS (2015) Frictional contact stress analysis of spur gear by using finite element method. Appl Mech Mater 772:159–163. https://doi.org/10.4028/www.scientific.net/AMM.772.159
16. Jani SS, Shah JB (2017) Design, modelling and analysis of helical gear pair using ANSYS and AGMA standards for calculating a bending and contact stress on gear profiles. International Journal of Advance Research and Innovative Ideas in Education 3(2):3813–3821

17. Jadhav PP, Bhaskar SV (2016) Design and analysis of helical gear made of stainless steel and nylon under different loading conditions. International Journal of Engineering Research and Technology 5(10):546–552
18. Lad V, Singh LP (2016) Design modelling and analysis of helical gear using CATIA, ANSYS and AGMA. International Journal of Mechanical Engineering and Technology 7(4):221–226
19. Thu MP, Min NL (2018) Stress analysis on spur gears using ANSYS workbench 16.0. International Journal of Science and Engineering Applications 7(8):208–213
20. Ghosh S, Ghosh R, Patel B, Srivastava T, Barman RN (2016) Structural analysis of spur gear using ANSYS workbench 14.5. Int J Mechanical Eng Technol 7:132–141
21. Fan XF, Zuo MJ (2006) Fault diagnosis of machines based on D–S evidence theory. Part 2: application of the improved D-S evidence theory in gearbox fault diagnosis. Pattern Recogn Lett 27:377–385
22. Altaf S, Mehmood MS, Imran M (2018) Implementation of efficient artificial neural network data fusion classification technique for induction motor fault detection. Journal of Engineering Sciences 5(2):E16–E21. https://doi.org/10.21272/jes.2018.5(2).e4
23. Pylypaka, S., Nesvidomin, V., Zaharova, T., Pavlenko, O., Klendiy, M.: The investigation of particle movement on a helical surface. In: Ivanov V. et al. (eds) Advances in design, simulation and manufacturing II. DSMIE 2019. Lecture notes in mechanical engineering. Springer, Cham, pp. 671–681 (2020) doi:https://doi.org/10.1007/978-3-030-22365-6_67
24. Tarelnik VB, Paustovskii AV, Tkachenko YG, Martsinkovskii VS, Belous AV, Konoplyanchenko EV, Gaponova OP (2018) Electrospark graphite alloying of steel surfaces: technology, properties, and application. Surf Eng Appl Electrochem 54(2):147–156. https://doi.org/10.3103/S106837551802014X
25. Konoplianchenko I, Tarelnyk V, Antoszewski B, Martsynkovskyy V, Belous A, Gerasimenko V, Vasilenko O (2018) Mathematical modeling a process of strengthening steel part working surfaces at carburizing thereof by electroerosive alloying method. In: 13th international conference Electromachining 2018, EM 2018, AIP conference proceedings, vol 2017, p 020008. https://doi.org/10.1063/1.5056271
26. Lishchenko NV, Larshin VP (2018) Profile gear grinding temperature reduction and equalization. Journal of Engineering Sciences 5(1):A1–A7. https://doi.org/10.21272/jes.2018.5(1).a1
27. Khovanskyi S, Pavlenko I, Pitel J, Mizakova J, Ochowiak M, Grechka I (2019) Solving the coupled aerodynamic and thermal problem for modeling the air distribution devices with perforated plates. Energies 12(18):3488. https://doi.org/10.3390/en12183488
28. Rogovyi A (2018) Energy performances of the vortex chamber supercharger. Energy 163:52–60. https://doi.org/10.1016/j.energy.2018.08.075
29. Merzliakov I, Pavlenko I, Chekh O, Sharapov S, Ivanov V (2020) Mathematical modeling of operating process and technological features for designing the vortex type liquid-vapor jet apparatus. In: Ivanov V et al (eds) Advances in design, simulation and manufacturing II. DSMIE 2019, Lecture Notes in Mechanical Engineering. Springer, Cham, pp 613–622. https://doi.org/10.1007/978-3-030-22365-6_61
30. Krol O, Sokolov V (2020) Research of modified gear drive for multioperational machine with increased load capacity. Diagnostyka 21(3):87–93. https://doi.org/10.29354/diag/126026
31. Krol O, Sokolov V (2020) Research of toothed belt transmission with arched teeth. Diagnostyka 21(4):15–22. https://doi.org/10.29354/diag/127193
32. Krol O, Sokolov V, Tsankov P (2020) Modeling of vertical spindle head for machining center. J Phys Conf Ser 1553:012012. https://doi.org/10.1088/1742-6596/1553/1/012012

Chapter 14
Security and Privacy of Cloud-Based Online Online Social Media: A Survey

Uma Shree Yadav, Brij B. Gupta, Dragan Peraković, Francisco José García Peñalvo, and Ivan Cvitić

1 Introduction

Social media [34, 40, 43, 44] sites including Twitter [35, 48, 54], Myspace, Instagram, Facebook, and Telegram have become very popular in the last several years [1]. Connectivity across social media platforms has brought people all together in a manner that they've become a part of everyday life. Many authors have described social networking websites as web-based applications that help users to make profiles within these sites, add a list of other users with whom he or she would like to link, and view or search their list of relations and all those created by others within the framework. Social media sites therefore provide a place for sharing and exchanging information between users [5, 11, 55].

Popular social media platforms like Instagram and Snapchat have thousands of millions or maybe even trillions of users spread across the globe exchanging shared content. Widely dispersed users worldwide could exchange data from various sources, several like really wide photos and videos, among each other. The proportion of these kinds of data items are rising significantly nearly every day. Users expect their social service provider to have low bandwidth exposure, consistency of data, reliability, and confidentiality requirements. Even so, social network [57, 84]

U. S. Yadav · B. B. Gupta (✉)
National Institute of Technology Kurukshetra, Kurukshetra, India

D. Peraković · I. Cvitić
University of Zagreb, Zagreb, Croatia
e-mail: dperakovic@fpz.unizg.hr; icvitic@fpz.unizg.hr

F. J. G. Peñalvo
University of Salamanca, Salamanca, Spain
e-mail: fgarcia@usal.es

L. Knapcikova et al. (eds.), *Sustainable Management of Manufacturing Systems in Industry 4.0*, EAI/Springer Innovations in Communication and Computing,
https://doi.org/10.1007/978-3-030-90462-3_14

service providers have restricted the amount of data management and storing resources to data access latency [67, 77, 78].

Data generated by online social media is very large in scale and maintaining such a wide scale data is not an easy task because user's dataset needs to be updated on a regular basis. So private data center maintenance could be unfeasible and unaffordable [36]. Barter exists between latency cost and storage of data cost; therefore, having a private data center for storing social media data is very expensive and is not easily affordable. Keeping this limitation of private data center in mind, we move toward a more appropriate solution, i.e., to use cloud data [42] centers which are more reliable and affordable. Cloud-based data center is less in cost as compared to private data center, which is easily affordable and provides many remarkable resources at not so high cost. Services provided by cloud-based data centers include storage, platform services, and maintenance services. These distributed cloud-based platforms have nearly infinite storing capacity and are ideal for wide scale data storage on online social media.

Basically, cloud computing [2, 45, 46, 79] acts as a service provider; it provides various services to the Internet users. It [87, 122, 123] simply provides computing services like database [110], servers, storage, networking, analytics, and applications. Cloud-based services provide flexible services and users have to pay only for the service they want to access at a time and on rent basis services are taken by the users. These services are easily available and storage and maintenance costs are not so high. Usually, users charge just for cloud services [80, 86] that they use, effectively reducing users' operational expenses, operate networks quite effectively, and expand once business requires growing.

Cloud computing [6, 38, 85, 88] comes into picture because social media has large amounts of data which is large in size and also more shareable in nature, so to handle such amount of data we need infrastructure which has great storage and maintenance capacity. Online social media platforms heavily rely on multimedia content like videos, images, post, photos, etc. Any uploaded content on social media gains popularity very quickly and soon after uploading people like and share such contents [9]. It could be videos of photos related to some popular events. Their sharing increases its volume and by this the data size increases tremendously. However, social media sites with cloud computing [27, 28, 76, 90] facilitate the storage capacity so that user's data can be easily managed and stored. Also, cloud-based social media provides various tools and services such as detection of keywords, tracking and prioritization of incoming posts on industry social media pages and comprehensive metrics, and social media success statistics [131, 132].

Cloud-based online social media having cloud services, utilized all these services efficiently to deeper comprehend the world and evolve in different fields of individual endeavor; the exponential rise of social media content enhances the risk of data security. There are various security issues in cloud-based social media, which is shown below [126–130]:

- Loss of data: Cloud services are expected to lose data from their users because of power failures.

- Risk from insiders: Sky-high Networks' alarming findings showed that around 1 in 10 business customers had their credentials mostly on Dark web [56, 58] for auction. The consequences may be immense if critical social media information gets out of control. Because cloud is remotely accessible, this problem is compounded as its difficult to precisely identify who signs into a hacked profile.
- Loss of oversight: On storing data in the cloud, online social service providers completely rely on cloud service providers for maintenance and give complete control of data to the cloud along with privacy and security of data.
- Nonconformity: Crimes linked to nonconformity toward rules and regulations will result in massive penalties for businesses. It ultimately decides to associate a third-party regarding confidential information that is quite challenging.

This study outlines a lot of research work undertaken to overcome various security and privacy problems in cloud-based online social media. Also includes taxonomy of security and privacy issues in cloud-based social media. This survey paper explores the privacy problems resulting from the use of cloud services on social media as well as suggests potential solutions to these problems.

2 History and Background

The success of any modern technology related to security is dependent upon the layer of security it provides. There are many research articles presented by researchers regarding cloud computing [99–103] and social media. As with the enhancement in Internet users, security concern also rises. Security of online data is a more concerned and important area; data generated by online social media needs to be secure, as the data is very large in volume and maintaining data in private databases is not possible so cloud-based services [2, 59–61] are now used to maintain huge amounts of data. By using cloud services storage problems have been solved to a great extent but still there are security and privacy issues. Also, with the rapidly emerging existence of social media and cloud services, clients are trying to experience new opportunities to engage with and leverage such evolving methodologies.

2.1 Overview of Cloud Computing

A very general definition of cloud computing [62, 64] is a collection of multiple connected servers which offers on-demand services such as storage, computing power, analytics, business applications as well as other tools, and technologies over the Internet. The large volume of data is stored on the cloud [63], this huge data is managed by a third party, so there is a risk of data security and confidentiality that data can be exploited or lost in the cloud system. Despite all of this, the cloud is flexible, simple to measure, and easy to manage at an affordable price [4121–125].

Mainly cloud computing is act as service [2, 6] providers it provides various services to the Internet users, it is simply provides computing services like database, servers, storage, networking, analytics and applications cloud based services provides flexible services and user have to pay only for the service he wants to access at a time, on the rent basis services is taken by the users. These services are easily available and storage and maintenance costs are not so high. Usually, users charge just for cloud services that he uses, effectively reducing users' operational expenses, operate networks quite effectively and expand once business requires to grow.

Cloud computing tackles a wide range of traditional challenges on hypothetical software creation and implementations in which you have a platform with scarce resources. These are some of the benefits of cloud computing: speed, performance, cost, global scale, productivity, reliability, performance, and security. In order to make cloud services easily available and feasible various services, resources, and models are included in cloud infrastructure.

Mainly there are two models of cloud computing, the first one is a service model and the second one is a deployment model. Deployment model consists of four deployment models such as public cloud, private cloud, community, and hybrid [89].

Public Cloud This cloud is public and is one of the most common clouds. Starting to use public cloud tools and services is simple for everyone, whether it is used by an individual or an organization. There is really no major investment spending up ahead. Users do not even have to influence buying devices or think regarding cloud configuration and maintenance. That is because the cloud providers procure and maintain all the resources, that is, the physical servers, computing, connectivity, etc. A cloud service company establishes and manages the cloud. Examples of such cloud types are Microsoft Azure and AWS [4]. On the net, we connect public cloud infrastructure and services. So, we need Internet connectivity in order to use a public cloud, and the cloud provider has a website interface that enables the cloud services and resources. We pay a subscription to the cloud service provider for the cloud tools and services we consume. This type of cloud includes sharing the similar resources such computers, tools, networks, and storage. Figure 14.1 shows the benefits, limitations, and use case of public cloud.

Private Cloud A private cloud is private to an entity or we can say services, tools, and resources of a private cloud are dedicated to an entity or a group. With the exception of the public cloud, in private cloud, cloud services are not sharable with other entities both software and hardware technologies or related resources and are devoted to one entity only. It can be served by a third-party provider. The key concept to remember is that it is private, which means all infrastructure (hardware and software) is devoted to only one organization. It's because the company controls all that is machinery, applications, and service. So, cloud users have full autonomy and can modify it in whatever manner they would like to fulfill specifications. In particular, private clouds are mostly used by federal agencies, commercial banks, or any large companies having corporate functions. Figure 14.2 shows the benefits, limitations, and use case of private cloud.

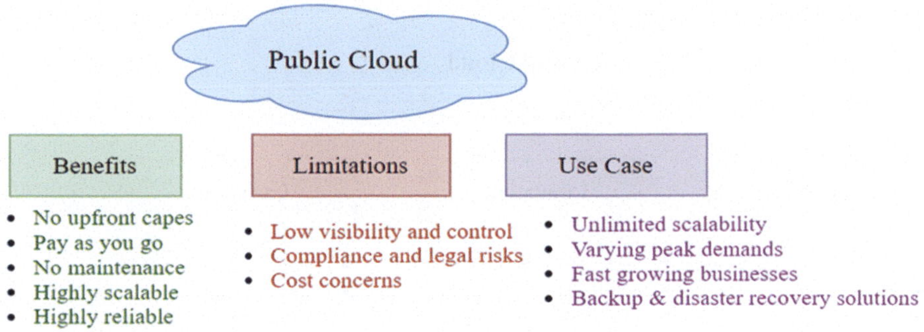

Fig. 14.1 Benefits, limitations, and use case of public cloud

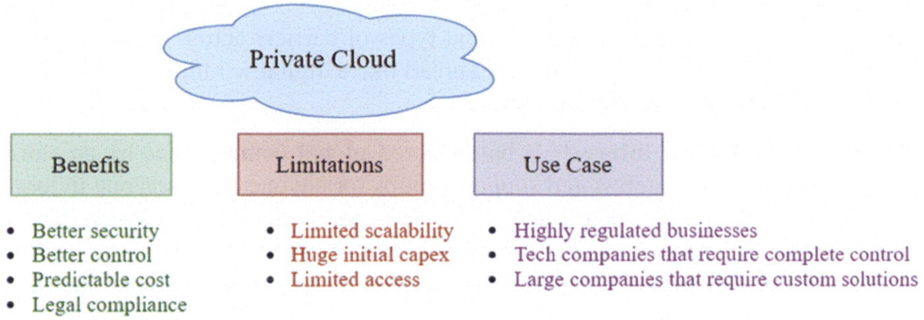

Fig. 14.2 Benefits, limitations, and use case of private cloud

Hybrid Cloud A hybrid cloud is a hybrid of private and public cloud; it offers benefits of both the clouds. It is the best combination of both the clouds; one can use this cloud with benefits of both worlds. Cloud bursting is a key concept offered by hybrid cloud. Figure 14.3 shows the benefits, limitations, and use case of hybrid cloud.

2.2 Overview of Online Social Media

Increasing use of social media has allowed people to share their huge content such as images and videos. The proper thing for users to relate and communicate across the globe is via the social media platform. Social media covers many other digital tools, even email and SMS are social media tools. Social media provides new ways to connect, interact, and learn. It is a platform where we can find people with similar interests, share knowledge and experiences, learn from each other, and have an opportunity to develop ourselves personally and professionally.

Social media has undoubtedly changed the world we live in today, it has made the world hyperconnected. It essentially revolutionized how we communicate and

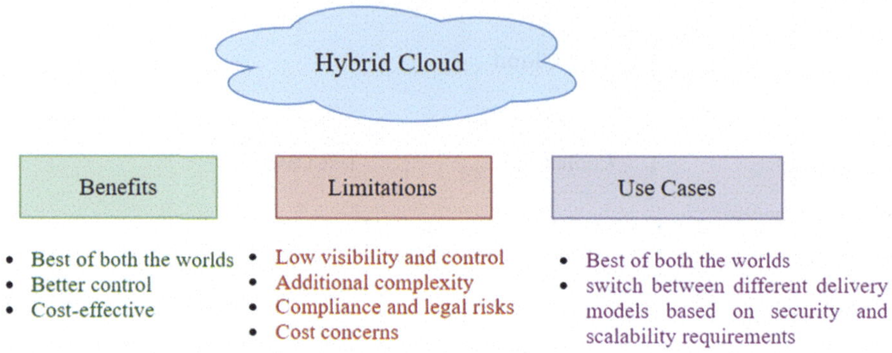

Fig. 14.3 Benefits, limitations, and use case of private cloud

connect as a global society where different types of barriers across countries have become blurred so information can now spread like a digital wildfire.

Benefits of online social media include:

- **Common classes of interest:** It has fostered global communities by creating common classes of interest and support groups for anyone and everyone to join.
- **Reconnect**: Social media has helped so many people locate long lost relatives and has even reunited adoptees with their birth parents.
- **Brands and consumers:** Consumers have found that they have a voice, a share in the conversation about brands.
- **Government and citizen interaction:** Governments worldwide are realizing that their traditional systems of information manipulation are no longer working and they have been forced to realign their strategy of communication with their citizens.
- **Democratic Potential:** Social media use is highlighting both the medium's democratic potential as well as its potential to undermine democracy.
- **Viral campaigns:** Some campaigns have raised millions of dollars for charities like The Ice Bucket Challenge.

Along with the abovementioned benefits of online social media there are many social media threats, [81] these are:

- **Trolls:** Through initiating disagreements or disturbing individuals by uploading offensive and off comments inside a social network [47, 49–51], troll is described as causing strife on the Network. Social networking [52] trolls are simply anybody who intentionally posts anything inflammatory to try to have a boost from several other clients.
- **Bots and bogus accounts:** Bogus accounts enhance active systems, as comments, likes, thoughts, also the Internet survey findings become distorted. Bots misrepresent the information which is used for estimation of information extraction, customer engagement, and customer behavior. Botnet attackers execute password stacking attempts then seize off profiles using compromised information.

- **Government repression:** The ability of social sites to spy on and regulate its own users, manipulate or affect the outcomes of polls, censor critical viewpoints, and eventually destroy online rights; and security is being leveraged by autocratic regimes and political misinformation purveyors across the globe.
- **Fake trends:** Many posts from paid trolls, bots, and fake accounts are not expressions coming from the people but are instead made-up to show hate, to magnify popularity of partisan views, or to influence others.
- **Misinformation:** Social media algorithms effectively decide which information a person is exposed to.
- **Viral campaign:** Because of our need as a society to share, information can be spread very quickly via social media. So with each share of a story, information can become more and more fabricated. Social media is a powerful tool that can be and is increasingly being used, to push agendas that can subvert the rule or law, create panic and social disruption, and violence. Many anonymous influencers have the tools or can get access to skills that create video clips, where unrelated footage is cleverly juxtaposed with voice-overs to create false documentaries to justify agendas.
- **Bullying:** Social media has also become a bully's most valuable tool. Cyber bullying [39] can destroy an individual's reputation and life with the click of a button.
- **Stalking:** Social media also opens the door for stalking.
- **Unregulated:** Social media is completely unregulated and often anonymous. Many mimics regulated media to push fake news or to access public figures of crimes without offering proof or evidence.

2.3 *Social Media Using Cloud*

In numerous roles, particularly private and commercial, the young group has endorsed the usage of social media [5]. Cloud computing was already optimized for the control of these heavy traffic. In the accessibility of the Internet, social media sites have assisted to a greater degree. Throughout the cloud storage systems, broad digital data is stored. Typically, a ton of bandwidth is filled with images, videos clips, and photos that are among the most commonly shared content on social media. Throughout the benefits rendered, including analytics and service quality regulation [6], cloud computing vendors such as salesforce and Amazon further consolidated. Consequently, clients can use these resources despite actually buying them, although they are hosted inside a cloud environment.

Social cloud is mainly used for storing huge amounts of data which is frequently generated by social media platforms. Using encrypted methodology on social media may give highly secured bandwidth inside the cloud; these services are now affordable at a lower price. Internet users are increasing very rapidly; along with this the social media sites are gaining lots of popularity with the ease of access. So the management of a cloud service provider extends the cloud infrastructure.

Cloud computing provides many services along with data storage, and social clouds are often used by social media for numerous different activities such as data mining and data analytics. Another advantage of cloud-based social media is that it offers simple access to a massive quantity of data. The social cloud provides authorization, authentication [105], and system administration functionality with social media. Many organizations leverage cloud services using various resources and functionality of cloud computing via social media platforms in order to perform various activities; nevertheless, these organizations face lots of difficulties and have the responsibility to secure users' personal dataset along with other confidential info.

Social media users can now leverage services or resources (shared) from everywhere in the world via cloud technology. That's also advantageous for social media because cloud technology retains and protects their customer's confidential data and always makes sure to keep data protected from any kind of security, although it is insignificant. The online social media acts like a community hub for peers, whereby mates, friends, and groups are formed and controlled. Every time when social media users share any new content with their friends, users basically use a private cloud for this and then information management application uses the framework of social media cloud to retrieve its database of social media friends, connections, and groups. With all these steps users set authorization to restrict data or content they have uploaded to a particular friend or connection.

3 Related Work

Different researchers have already addressed security and privacy issues in cloud environments with social media. So, this section basically gives an overview of related work on security and privacy in cloud-based social media.

Author Deyan Chen et al. [8] describe the standards for privacy and security in cloud-based social media and along with this they also describe the criteria or conditions for safe sharing of data over cloud.

Author Seyed et al. [9] describe the various cloud service issues such as privacy, integrity, accessibility, transparency, and anonymity, and extensively discusses the challenges of all of the issues and protection mechanisms. Author Wang et al. used pilot testing to describe the security and privacy enforcement of SaaS infrastructure and also includes various security and privacy threats of cloud.

In cloud technology, Wang et al. [10] examined variables that impact the management of information security. It describes the required security criteria for organizations to grasp the complexities of cloud data protection.

Author Meiko [11] addresses the technological safety problems relating to the application of the cloud environment, like Xml exploits, threats linked to plugins, and breaching threats. Cloud platform security breaches that exist inside the cloud and also the study classifying the existing weaknesses onto digital, cloud-related, access-related, and vulnerability assessment features are described by Bernd [12].

Paper presented by Subhashini concentrates on a system as service infrastructure and also addresses the security model of cloud services. Cloud security [25, 26, 82] alliances address crucial issues of cloud computing at international conferences. They discussed a number of excellent policies related to cloud security [113–118] and policies.

A systematic overview of the protection of advanced analytics is provided by Tran [13]. To provide comprehensive viewpoints and thorough characterization, the researchers initially present the possibly best taxonomy, then provide perspectives toward early projects, and eventually define available areas for future research.

Author Youke et al. [14] are researching the risk safety approach in social media focusing on microbial scenario estimation. In order to protect toward security attacks, the study identifies the boundary points of threat and risk amplification and suggests a threat restriction approach based on distribution estimation.

Authors Xin and Yizhu [15] consider the issue of provable subject level quest throughout the procurement paradigm of social media data and suggest two methods to ensure the validity of social media data. Data consumers use certain obtained search queries and validation artefacts in the effective part to empirically check the accuracy of sensitive attributes. Throughout the intensified paradigm, to minimize the amount of deemed value, social platforms will organize all issues in k-clusters.

Author Mowbray presented a client-based confidentiality platform [17] throughout the storage of data and usage phases. In order to support people, monitoring the storage [106] and confidential data present in the cloud offers a user-centered security framework. Mulero et al. addressed the difficulties posed by established security technology systems for example, k-anonymous, anonymization of graph [32], and data pre-processing methods; whenever related to huge amounts of data and evaluated available solutions [16]. The task of data confidentiality is to disclose effective privacy knowledge about individual rights. Also, author Randike suggested a system for the security of confidentiality focused on attributes of data transparency [18]. Customers that obtain information as well as the pieces of data they use could be detected by an information accountability operator. The operator describes a number of ways to hold clients responsible for abuse if improper usage is observed.

Research reports showed that cloud infrastructure is quite vulnerable to privacy and security threats and lots of significant research has been done on that area for reducing the privacy and security problem.

4 Privacy and Security Issues

The greatest hurdle that prevents businesses from joining the clouds is confidentiality. In industry-related programs, protection is indeed concentrated. Data protection is extremely difficult to measure and sometimes measurably equate, and in contrast to other characteristics in technical contexts. For this purpose, the risk analysis of cloud services [17] should largely rely upon the credibility of the business and, finally, true past record. Although it is harder to identify such authentic past records among businesses, cyberattacks might not be reported by the media until needed by law.

Industries who use cloud storage would like to assure that their info, content, and data are protected from potential and domestic hackers. Data breaches and sniffing are counteracted by storing data in encrypted form over the cloud, although cryptography is unable to avoid Distributed of Denial of Service (DDoS) attacks [19, 65, 111, 137, 138] to prevent data manipulation and destruction.

4.1 Privacy and Security Issues Faced by Cloud

The concept of data security in the cloud [112] is very close to the concept of conventional security and confidentiality of content [8, 24]. Also, there is a complexity in cloud-based security and privacy because the cloud has various characteristics such as transparency and most importantly multiuser and sharing. There'll be a great deal to do with it if a conversation is held regarding cloud protection. The cloud providers guarantee that the user does not really experience any issue [20], like losing data and private information; however, there is a risk that a cybercriminal, who can imitate a genuine user, will infiltrate the cloud by harming the whole cloud. This contributes to the effect of multiple clients sharing the contaminated cloud.

There are different problems associated with cloud systems [17, 20]. The biggest worries about privacy and security involve:

- **Multi-tenancy:** Problems like deploying applications, data privacy, data access, or availability requirements must be factored to create a stable multiuser and sharing framework [19].
- **Heterogeneousness:** Heterogeneity characterizes cloud systems. Multiple service providers have various methods to implement privacy and security protections that lead to obstacles in adoption of the system [20].
- **Parallelization:** Generally, every data available on cloud storage is digitally obtained. A reasonable framework must be aimed at improving safety and confidentiality required to preserve effective autonomy or facilitate contact and collaboration among virtual devices [22]. To implement the accessibility standards to protect how compute nodes exchange content over cloud server, an access management scheme that is very versatile could be used.
- **Outsourcing:** Online social clients suffer losses of their own data with cloud computing. Throughout this strange situation confidentiality becomes a major issue. Thus, it is necessary for cloud providers to implement an effective procedure which prohibits user data from being used for reasons besides those intended [21].

Main challenges of cloud infrastructures are confidentiality, availability, and integrity. Table 14.1 also shows the various challenges to security and privacy

Table 14.1 Challenges of privacy and security issues

Categories	Problems
Security issues	1. Trust issue 2. Access issue 3. Audit issue 4. Availability issue 5. Multi tenancy issue
Privacy issues	1. Misuse of cloud computing 2. Malicious insiders 3. Trans border data flow 4. Dynamic provision

issues. Confidentiality includes the privacy of the information on machines which are linked to the Internet. The first is that the National Security Agency (NSA) has a huge Online connectivity spy service, including a 3-day recurring database [107] engine. The risks involved in data analysis of centralized control have gained devastating exposure mostly on the verge of Framework utilizing classified monitoring systems. The credibility of computers on the Network could be abused by organizations like the NSA. There are several additional threats possible by several other people or organizations like modifications into the application of administration or account hacking. There are a few problems with regard to availability. The demand for cloud storage is one among them. Amazon, for instance, experienced major outages, causing damage to AWS clients. Hacktivists assaults, supplier unintentional objectification, natural catastrophic events such as fires or disasters, and suppliers can reflect aspects wherein cloud users can experience irreversible data loss [128].

4.2 Risks to Cloud Security

Many kinds of risks related to privacy and security that cloud is susceptible to are shown in Table 14.2; it gives a detailed summary of the risks (threats) to users (cloud users) as per the security triad which is basically a security model that includes CIA (confidentiality, availability, and integrity), which is important to every cloud model that offers cloud services.

4.3 Security and Privacy Issue in Online Social Media

Users of online social networking sites have been subjected to security issues due to the increased use of online social networking sites; these issues can be phishing attack, SQL injection, DDoS attack [121, 124, 125, 136], cross-site scripting, drive by download, and many more [7, 10]. There are various categories of privacy and security issues which are shown in Table 14.3.

Table 14.2 Security risks to the cloud

Category	Threats
Confidentiality	Spillage of information: • Access control authority deficiency throughout many contexts. • Inability of internal and external transmission systems for cloud data and updates. Risks to internal users: • Malevolent customer of a cloud service. • User of compromised cloud clients. • Deceptive customers of third parties. Spillage of information: • Access control authority deficiency throughout many contexts. • Inability of internal and external transmission systems for cloud data and updates. Risks to internal users: • Malevolent customer of a cloud service. • User of compromised cloud clients.
Availability	Risks from outside attackers: • Remote attacks on cloud platform. • Attacks on cloud applications by using software remotely. • Attacks on cloud hardware remotely. • Extortion of customers of cloud providers and customers of cloud clients.
Integrity	Quality of data: • Overview of defective client applications or operational environments. Repression of data: • Improper specification of virtual and hypervisors platforms. • Security boundary lines described wrongly. Access by customer: • Inadequate techniques for authentication and authorization.

Table 14.3 Categories of privacy and security issues of social media

Attacks on Online social media	Descriptions
Spam [29, 30, 72] and malware	Media spam [73, 75] is undesirable spam data that occurs on social media platforms [31], online forums as well as any webpage containing users' data such as chat messages. Because of the inherent interpersonal trust between internet friends, media spam spreads like wildfire through online environments that inspires a user to open that message or lure them to click on links shared by their mates.
Exploited account	Exploited accounts are genuine user accounts that were already built and were used by original rightful holders and had been exploited by hackers.
DDoS	DDoS [92, 133–135] is an attack emerging from a specific source and several origins are being used to initiate the attack concurrently. Deploying this form of attack floods the objective with requirements, thereby overburdening the machine.
Sybil attacks [53]	Sybil attacks are defined as a person using fake personalities to control the output of a service [33]. Even though online social media has become susceptible to Sybil attacks by regulating several profiles, Sybil users are unlawfully growing their control and authority in online social media [37].

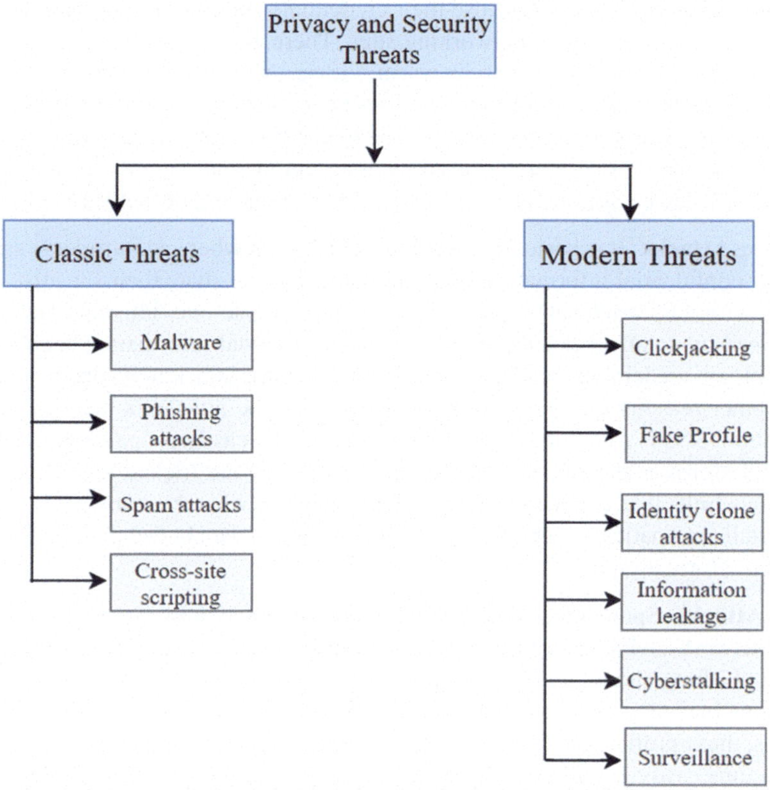

Fig. 14.4 Classification of privacy and security threats in OSNs

Over a span of time, social media platforms become an essential and important part of society especially for youngsters, with this security concern also growing. Some of the security issues may include viral ads, malware attacks, network attacks, phishing attacks, clickjacking, and many more that come under classic and modern threat. Figure 14.4. is showing classification of online social networks threats [66, 68, 69].

- Classic threats affect users of Online Social Network (OSNs) and other online users – not necessarily the user of any social networking sites. Threats include spam malware [4], phishing [3], spam [2], and cross-site scripting attacks [5].
- Whereas modern threats are the threats that only make social networking site users vulnerable.

Malware Malware is a malicious intent software which is typically used by a hacker to inflict significant data and device harm or to obtain unauthorized remote access. The framework of OSNs and their connectivity feature between people make it a lot easier to target malware on networking sites compared to competing services. Malware has the ability to get private information of users by manipulating and impersonating them to disclose their credentials and also to lure users to share malicious content via social networking sites. There are various types of malwares with different functionality, such as worm, viruses, trojan horse, ransomware, etc. Attacker uses them according to the actions he wants to perform on a user's computer with the intent to access users' credentials. The koobface malware is a malware found in Facebook which is used by attackers to obtain login information and make attack device bot for further propagation or spreading of malware [119].

Phishing Attacks It is basically a kind of online scam where an intruder is seeking to pilfer confidential information by tricking the users to share their sensitive information with the intruder who pretends to be a legitimate one. Phishing attacks are performed by an attacker with an intent to steal personal information such as card details, login credentials, bank account details, and this attack may also be used for many other reasons that completely depends upon the attacker's needs. Phishing attack on OSNs is to get login credentials and can be easily propagated via online OSNs platform as their nature of sharing. Victim is often tricked into clicking on malicious links that can lead into malicious websites whereby malware download and install automatically, which may eventually causes data breach for stealing confidential information [6].

Spam Attacks Spam attacks include sending unsolicited messages by using online social networks, which basically involves commercial advertising. Spam attack may lead to phishing attack; its ultimate aim is to steal confidential data and it could be malicious message, post, content, and link. Spam messages can be spread by fake profiles, these profiles are formed by an attacker with the name of a popular face so that victims easily get into trap by believing the post posted by these profiles. So it is obvious that fake messages [91] are mostly spread by a hacked profile [7]. Spam filtering [8] techniques are used for filtering the spam-containing content.

Cross-Site Scripting It is abbreviated as XSS [74, 95, 104], it is an attack on security which is done by injecting malicious code on the client side by exploiting web-based vulnerability. In this attack, hacker uses web services to transit malicious software [9, 108, 109] and intruder injects malicious scripts into web pages which are accessed by others, and clicking or accessing such kind of web pages may cause serious security problems such as data theft from cookies, identity theft, and stealing login credentials, also social networking sites can be used as medium for their propagation in the form of worms that may propagate and contribute in cyber-attack.

Clickjacking Clickjacking is also called UI redressing, in this attack; a hacker gives a website to the user that looks legit and the user is lured into clicking it in the interface but in reality, the legit-looking website containing a malicious designed page is created by an intruder to exploit the user's confidentiality. In this attack an intruder can trick users to upload malware containing posts on their social media [70], [71] profile and lure them to click on posts for liking and commenting so that particular post gets shared among their friends; when an attacker is successful with his intention, he can use the user's computer such as camera and mic to keep eye on user's activities.

Fake Profile Usually a fake profile is created in order to perform a malicious act. It is very easy to create a fake profile by fake name, it could be any person, business, or organization which is used to create fake users. Sometimes attackers use popular names or face as a bait to a social media user to deceive or to exploit data. In order to spread malware, multiple requests are sent to legitimate users and somehow these legitimate users become a carrier of spams; these spasms cause security breaches like exploiting private information, and also monitor activities of users that can be further used to perform other major attacks [97].

Information Leakage Social networking is mostly about accessing and sharing data freely with peers. Many users share their personal data on a voluntary basis, like wellness data. Even worse, too much private details about goods, projects, organizations, or some other sort of sensitive information is exchanged by some of them. It could have adverse implications for Social media users to share these sensitive and confidential information.

Cyberstalking Cyberstalking is used via social media and online networking to annoy a person or community. It may be used for surveillance, stealing of identification, assaults, sexual solicitation, or harassment. Cyberstalking can vary in appearance, as described, but it is stalking or assault, which actually occurs via digital platforms like social media, websites, or messages in the wider definition. Cyberstalking may not require direct contact to be involved, and some victims do not even know that they've been stalked online. Via different approaches, criminals can track victims and use the information collected for crimes such as identity fraud [37]. The distinction among the cyber world and actual life could become fuzzy in certain instances. Hackers will gather confidential info from yourself, reach your associates, even try to threaten you offline.

Surveillance Surveillance of social networks is indeed a modern form of oversight that differs from a person's personal socializing and social positions throughout politics, the economy, and society organizations. This becomes a control mechanism for utilizing their accounts and connections with others the various characteristics of their users in various social roles. Social networking monitoring is an innovation surveillance scheme for the tracking of individual activity on social media.

4.4 Solutions to Threats for Data Protection

A data privacy aspect influences cryptography method, authorization, authentication, client's data security, and privacy; retrieval must be built by cloud storage service vendors to improve the protection of crowded data [41, 23, 83]. In order to

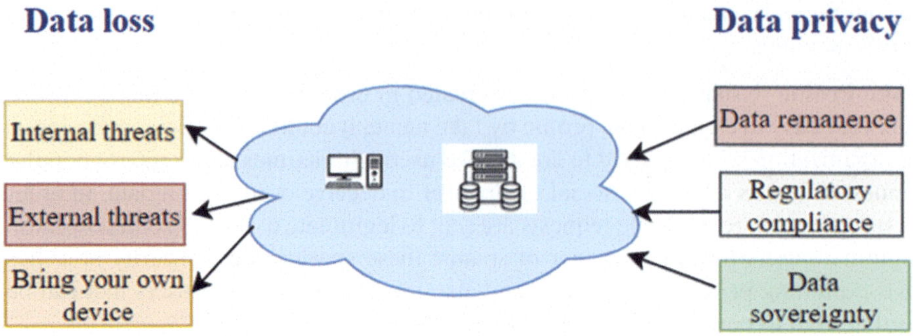

Fig. 14.5 Protecting Cloud Data

Table 14.4 Crucial security problems and its potential protection measures

No.	Threats to privacy and security	Methods for protection
1	Identity impersonating	• Identity verification. • Secure personal information. • Do not let data be public.
2	Alteration of information	• Accreditation. • Security measures for interference resistance. • Verification standards for messages. • Digital identification.
3	Threats related to denial of service	• QoS. • Screening and filtering. • Identity verification and authorization.
4	Repudiation threats	• Audit log. • Digital identification.
5	Retraction of data	• Authorization. • Privacy-boosted protocols. • Encoding. • Private details secure. • Do not disclose private information on public platform.

ensure data protection on cloud [120] various measures have been taken and a few of them are shown in Fig. 14.5.

Confidentiality as a utility must be given such that data confidentiality is improved. Private and most crucial data should be encrypted during data sharing by customizing the secure socket layer. Safety must be given utmost importance inside a multiuser and sharing framework. Confidentiality must be guaranteed in specific context for each level contained within design of a cloud service. To promote the usage of cloud in online social media [24], the service provider must cope completely regarding legal damage. There are various security-related threats and their measures are described in Table 14.4.

In order to ensure accountability about the storing and analysis of information over the cloud, regulatory enforcement and risk mitigation must be well stated

Table 14.5 Benefits and drawbacks of current technologies for cloud computing

Existing solutions	Benefits	Drawbacks
Framework for protecting privacy	Preserving confidentiality	Offering descriptive access privileges for machines
Enduring database confidentiality	Provides anonymity and accessibility	Not really effective enough to accomplish all security problems
Client-based privacy manager	Preserving confidentiality	Require honest cooperation of service provider
Privacy control of retained access	Ensure data protection despite exposing	Data procurement and unreliable cloud server problems
Scheme relies on identity	Easy and scalable	Restricted in terms of services
Retaining things secret while processing	Effective for the protection of privacy and resilient for network delays.	Focuses only on nondeterministic polynomial time problems
Information or data and social accountability mechanisms for storage-security	Extremely effective, responsive, and protected	Requiring an operational third-party monitor
Securing the storage data using RC5	Secure and easy to use	Cracking algorithm is a huge problem
Identity verification of service driven personality	Preservation of data protection.	No multiservice assistance and no realistic design help
Protection uses elliptical curves cryptography	Preserving anonymity and identity verification	Strives to maintain all the safety concerns
Fog computing solution	Safeguards from information leakage	Not solving all the problems
Solution including FHE (fully homomorphic encryption) entirely homomorphic encryption.	Effective tool toward preserving security	No practical implementation
Protect the misuse of content from indexing	Provide protection of data from leakage	Major drawback is dependent on indexing
Client-based privacy management	Reduce the risk and give extra secrecy	Need sincere support from the service provider

inside the agreement between the client and the cloud service provider [25]. It could reinforce mutual faith between the two and also ensure the protection of their most private data for people on social media. In order to maintain data accessibility, it is very important for consumers to try holding up sensitive information locally. Table 14.5. shows the benefits and drawbacks of current technologies for cloud computing [93, 94, 96, 98].

Information encoding is one of the best appropriate methods to secure private records [26]. Prior to actually storing this in the remote server, consumers must encode their content. The authorized person may then determine the participants of a specific community who are permitted to obtain the data. Shared data managed by interconnected data-centric protection must be ensured [131, 132].

5 Conclusion and Future Research Directions

Cloud Computing continues to be a modern or changing model whereby computation is perceived as a service on demand. When the company makes the choice to migrate to the clouds, it loses sight over the details. Cloud security depends on reliable encryption and computation. Cloud computing technology gives consumers of social media quite a lot of issues in terms of data protection, despite the amazing advantages. The number of social media users is growing day by day, which is why interaction among social media platforms and cloud computing is increasing. As a consequence of this, emerging confidentiality and trusting related concerns have been found. Many Internet users are conscious of security issues associated with personal data and so as a consequence, initiatives to secure such data are regularly created. In preserving the confidentiality and protection of personal data, including social media consumers and cloud services, vendors play a significant role. Throughout this study, in order to avoid the threats related to cloud storage, confidentiality issues and potential solutions to security problems are presented.

References

1. Obar JA, Wildman S (2015) Social media definition and the governance challenge: an introduction to the special issue. Telecommun Policy 39:745–750
2. Mell P, Grance T (2011) The NIST definition of cloud computing (Draft), special publication 800–145 (Draft). National Institute of Standards and Technology, Gaithersburg
3. Annapoorani, P. Indira Priya, 2014 Inferring Private Information from Social Network Using Collective Classification, International Journal of Innovative Research in Computer and Communication Engineering, 4, 1,
4. Amazon Elastic Compute Cloud (Amazon EC2) (2014) Amazon, http://aws.amazon.com/ec2/, Amazon. Accessed in Sept 2014
5. Zhao X, Salehi N, Naranjit S, Alwaalan S, Voida S, Cosley D (2013) The many faces of Facebook: experiencing social media as performance, exhibition, and personal archive. In: Proceedings of the SIGCHI conference on human factors in computing systems. ACM, pp 1–10
6. Hugos MH, Hulitzky D (2010) Business in the cloud: what every business needs to know about cloud computing. Wiley, p 139
7. Divya R, Mahesh B, Ushasree R (2014) Data Implication Attacks on SocialNetworks with Data Sanitization. International Journal of Current Engineering and Technology 4(3)
8. Chen D, Zhao H (2012) Data security and privacy protection issues in cloud computing. In: International conference on computer science and electronics engineering
9. Ahmadinejad SH, Fong PWL (2013) On the feasibility of inference attacks by third-party extensions to social network systems. In: Proceedings of the 8th ACM SIGSAC symposium on information, computer and communications security. ACM
10. Pranay R, Pavan Kumar P (2014) A Survey on Obstruction of Confidential Information Attacks in Social Networks. International Journal of Research in Information Technology 2(6)
11. Zheleva E, Getoor L (2011) Privacy in social networks: a survey. In: Social network data analytics. Springer US, pp 277–306

12. Ford R (2011) Traian Marius Truta, and Alina Campan. In: P-Sensitive KAnonymity for social networks, pp 277–306
13. Hong-yen tran j H (2019) Privacy-persevering big data analytics a comprehensive survey, parallel distrib. Comput 134:207–218
14. Youke Wu H et al (2019) A risk defense method based on microscopic state prediction with partial information observations in social networks. J Parallel Distrib Comput 131:189–199
15. Xin yao Y z et al (2019) Topic-based rank search with verifiable social data outsourcing. J Parallel Distrib Comput 134:1–12
16. Muntés-Mulero V, Nin J (2009) Privacy and anonymization for very large datasets. In: Chen P (ed) Proceedigs of the ACM 18th international conference on information and knowledge management, CIKM 2009. Association for Computing Machinery, New York, pp 2117–2118
17. Bowers KD, Juels A, Oprea A (2009) Proofs of retrievability: Theory and implementation. In: Sion R (ed) Proceedings of the 2009 ACM Workshop on Cloud Computing Security, CCSW 2009, Co-Located with the 16thACM Computer and Communications Security Conference, CCS 2009. Association for Computing Machinery, New York, pp 43–54
18. Gajanayake R, Iannella R, Sahama T (2011) sharing with care an information accountability perspective. Internet Comput IEEE 15:31–38
19. Bojović P, Bašičević I, Ocovaj S, Popović M (2019) A practical approach to detection of distributed denial-of-service attacks using a hybrid detection method. Comput Electr Eng 73:84–96
20. Dillon T, Wu C, Chang E (2010) Cloud computing: issues and challenges. In: 2010 24th IEEE international conference on advanced information networking and applications. Ieee, pp 27–33
21. Eldewahi AE, Hassan A, Elbadawi K, Barry BI (2018) The analysis of MATE attack in SDN based on STRIDE model. In: Proceedings of the international conference on emerging internetworking, data and web technologies, pp 901–910
22. D.Song, E. Shi, I. Fischer, and U. Shankar, "Cloud data protection for the masses", 2012
23. Hugos MH, Hulitzky D (2010) Business in the cloud: what every business needs to know about cloud computing. Wiley, p 139
24. Robison WJ (2010) Free at what cost? Cloud computing privacy under the stored communications act. Georgetown Law J 98(4)
25. Vaidya M (2016) Handling critical issues of big data on cloud. Managing Big Data Cloud Comput Environ:100
26. Zissis D, Lekkas D (2012) Addressing cloud computing security issues. Futur Gener Comput Syst 28(3):583–592
27. Rittinghouse JW, Ransome JF (2016) Cloud computing: implementation, management, and security. CRC press
28. D.Song, E. Shi, I. Fischer, and U. Shankar, "Cloud data protection for the masses", 2012
29. Kreibich C, Kanich C, Levchenko K, Enright B, Voelker GM, Paxson V, Savage S (2009) Spamcraft: an inside look at spam campaign orchestration. In: Proceedings of the second USENIX workshop on large-scale exploits and emergent threats, LEET
30. Cresci S, Di Pietro R, Petrocchi M, Spognardi A, Tesconi M (2017) The paradigm-shift of social spambots: evidence, theories, and tools for the arms race, in. In: Proceedings of the twenty-sixth international conference on world wide web companion, international world wide web conferences steering committee, pp 963–972
31. Yu R, He X, Liu Y (2015) Glad: group anomaly detection in social media analysis. ACM Trans Knowl Discov Data (TKDD) 10(2):18
32. Jiang M, Cui P, Beutel A, Faloutsos C, Yang S (2016) Catching synchronized behavior in large networks: a graph mining approach. ACM Trans Knowl Discov Data (TKDD) 10(4):35
33. Meinig M, Sukmana MI, Torkura KA, Meinel CJPCS (2019) Holistic strategy-based threat model for organizations. Proc Comput Sci 151:100–107

34. He W, Liu X, Ren M (2011) Location cheating: a security challenge to location- based social network services. In: Proceedings of the thirty-first international conference on distributed computing systems (ICDCS). IEEE, pp 740–749
35. Chu Z (2012) Detecting social spam campaigns on twitter. In: Proceedings of the 2012 conference on applied cryptography and network security. Springer, Berlin, Heidelberg, pp 455–472
36. Singh HJ, Bawa S (2018) Scalable metadata management techniques for ultra-large distributed storage systems—a systematic review. ACM Comput Surv (CSUR) 51(4):82
37. Giatsoglou M, Chatzakou D, Shah N, Beutel A, Faloutsos C, Vakali A (2015) Nd-sync: detecting synchronized fraud activities. In: Proceedings of the 2015 Pacific-Asia conference on knowledge discovery and data mining. Springer, pp 201–214
38. Al-Nawasrah A, Almomani AA, Atawneh S, Alauthman M (2020) A survey of fast flux botnet detection with fast flux cloud computing. Int J Cloud Appl Comput (IJCAC) 10(3):17–53
39. Tounsi W, Rais HJC (2018) A survey on technical threat intelligence in the age of sophisticatedcyber-attacks. Comput Secur 72:212–233
40. Yu R, He X, Liu Y (2015) Glad: group anomaly detection in social media analysis. ACM Trans Knowl Discov Data (TKDD) 10(2):18
41. Viswanath B, Bashir MA, Zafar MB, Bouget S, Guha S, Gummadi KP, Kate A, Mislove A (2015) Strength in numbers: robust tamper detection in crowd computations. In: Proceedings of the 2015 ACM on conference on online social networks. ACM, pp 113–124
42. Tan Y, Wu F, Wu Q, Liao XJTJOS (2019) Resource stealing: a resource multiplexing method formix workloads in cloud system. J Supercomput 75(1):33–49
43. Stringhini G, Kruegel C, Vigna G (2010) Detecting spammers on social networks. In: Proceedings of the twenty-sixth annual computer security applications conference. ACM, pp 1–9
44. Kayes N, Kourtellis D, Quercia A, Iamnitchi FB (2015) The social world of content abusers in community question answering. In: Proceedings of the twenty-fourth international world wide web conference. ACM, pp 570–580
45. Badger L, Grance T, Patt-Corner R, Voas J (2011) Draft cloud computing synopsis and recommendations. National Institute of Standards and Technology (NIST) Special Publication 800-146. US Department of Commerce. May 2011. Available online at: http://csrc.nist.gov/publications/drafts/800- 146/Draft-NIST-SP800-146.pdf. Accessed on: 20 Nov 2012
46. Mishra P, Pilli ES, Varadharajan V, Tupakula U (2017) Intrusion detection techniques in cloud environment: a survey. J Netw Comput Appl 77:18–47
47. Egele M, Stringhini G, Kruegel C, Vigna G (2013) Compa: detecting compromised social network accounts. In: Proceedings of the 2013 symposium on network and distributed system security (NDSS)
48. Stringhini G, Wang M, Egele C, Kruegel G, Vigna H, Zheng BY, Zhao (2013) Follow the green: growth and dynamics in twitter follower markets. In: Proceedings of the 2013 conference on internet measurement conference, IMC '13. ACM, New York, pp 163–176
49. Wisniewski P, Knijnenburg B, Lipford HR (2017) Making privacy personal: pro- filing social network users to inform privacy education and nudging. Int J Hum Comput Stud 98:95–108
50. Cheng Y, Park J, Sandhu R (2013) Preserving user privacy from third-party ap- plications in online social networks. In: Proceedings of the twenty-second international conference on world wide web companion. ACM, pp 723–728
51. Xue J, Yang Z, Yang X, Wang X, Chen L, Dai Y (2015) VoteTrust: leveraging friend invitation graph to defend against social network Sybils. In: IEEE Infocom. IEEE, pp 2400–2408
52. Koll D, Schwarzmaier M, Li J, Li X-Y, X. (2017) Fu, thank you for being a friend: an attacker view on online-social-network-based Sybil defenses. In: Proceedings of the thirty-seventh IEEE international conference on distributed computing systems workshops (ICDCSW). IEEE, pp 157–162
53. Zhang J, Zhang R, Sun J, Zhang Y, Zhang C (2016) TrueTop: a Sybil-resilient system for user influence measurement on twitter. IEEE/ACM Trans Netw 24(5):2834–2846

54. Zangerle E, Specht G (2014) Sorry, I was hacked: a classification of compromised twitter accounts. In: Proceedings of the twenty-ninth annual ACM symposium on applied computing. ACM, pp 587–593
55. Ruan X, Wu Z, Wang H, Jajodia S (2016) Profiling online social behaviors for com- promised account detection. IEEE Trans Inf Forensics Secur 11(1):176–187
56. Mayer JR, Mitchell JC (2012) Third-party web tracking: policy and technology. In: Proceedings of the 2012 IEEE symposium on security and privacy. IEEE, pp 413–427
57. Krishnamurthy B (2013) Privacy and online social networks: can colourless green ideas sleep furiously? IEEE Secur Priv 11(3):14–20
58. Takano Y, Ohta S, Takahashi T, Ando R, Inoue T (2014) MindYourPrivacy: design and implementation of a visualization system for third-party web tracking. In: Proceedings of the twelfth international conference on privacy, security and trust, IEEE, pp 48–56
59. Wang C, Chow S et al (2013) Privacy-preserving public auditing for secure cloud storage. IEEE Trans Comput 62(2):362–375
60. Zhou M, Mu Y et al (2011) Privacy-preserved access control for cloud computing. In: International joint conference of IEEE TrustCom 2011/IEEE ICESS 2011/FCST 2011, pp 83–90
61. Tang J, Cui Y, Li Q, Ren K, Liu J, Buyya R (2016) Ensuring security and privacy preservation for cloud data services. ACM Comput Surv (CSUR) 49(1):13
62. Wang J, Zhao Y et al (2009) Providing privacy preserving in cloud computing. Int Conf Test Measur 2:213–216
63. Hong JB, Nhlabatsi A, Kim DS, Hussein A, Fetais N, Khan KMJCN (2019) Systematic identification of threats in the cloud: a survey. Comput Netw 150:46–69
64. Sharma P, Sood SK, Kaur S (2011) Security issues in cloud computing. In: Mantri A, Nandi S, Kumar G, Kumar S (eds) HPAGC 2011, CCIS, vol 169. Springer, Heidelberg, pp 36–45
65. Rai S, Sharma K, Dhakal D (2019) A survey on detection and mitigation of distributed denial-of service attack in named data networking. In: Sarma H, Borah S, Dutta N (eds) Advances in communication, cloud, and big data, Lecture notes in networks and systems, vol 31. Springer, Singapore
66. Annapoorani, P. Indira Priya, Inferring Private Information from Social Network Using Collective Classification, Int J Innovat Res Comput Commun Eng, 4, 1, 2014
67. Beach A, Gartrell M, Han R (2010) q-Anon: Rethinking anonymity for social networks. In: Proceedings of IEEE second international conference on social computing (SocialCom), Minneapolis, pp 185–192
68. Lan L, Jin SJH (2010) Anonymizing social network using bipartite graph. In: Proceedings of international conference on computational and information sciences (ICCIS), Chengdu, pp 993–996
69. Divya R, B. Mahesh and R. Ushasree, "Data implication attacks on social networks with data sanitization", Int J Curr Eng Technol, Vol. 4, No. 3 (2014)
70. Obar JA, Wildman S (2015) Social media definition and the governance challenge: an introduction to the special issue. Telecommun Policy 39:745–750
71. Shozi NA, Mtsweni J (2017) Big data privacy in social media sites. In Proceedings of the 2017 IST-AfricaWeekConference (IST-Africa), Windhoek, Namibia, Southern Africa, 30 May–2 June 2017. pp. 1–6
72. Gao H, Hu J, Wilson C, Li Z, Chen Y, Zhao BY (2010) Detecting and characterizing social spam campaigns. In Proceedings of the 10th ACM SIGCOMM conference on internet measurement, Melbourne, Australia,1–3 November 2010; pp. 35–47
73. Gao H, Chen Y, Lee K, Palsetia D, Choudhary AN (2012) Towards online spam filtering in social networks. In Proceedings of the 19th annual network & distributed system security symposium, San Diego, CA, USA,5–8 February 2012. pp. 1–16
74. Gupta S, Gupta BB (2017) Cross-site scripting (XSS) attacks and defense mechanisms: classification andstate-of-the-art. Int J Syst Assur Eng Manag 8:512–530

75. Thomas K, Grier C, Ma J, Paxson V, Song D (2011) Design and evaluation of a real-time URL spam filtering service. In 2011 IEEE symposium on security and privacy (pp. 447–462). IEEE
76. Mitra A (2019) On investigating energy stability for cellular automata based pagerank validation model in green cloud. Int J Cloud Appl Comp (IJCAC) 9(4):66–85
77. Gulyás GG, Simon B, Imre S (2016) An efficient and robust social network De-anonymization attack. In: Proceedings of the workshop on privacy in the electronic society, Vienna, Austria, pp 1–11
78. Heatherly R, Kantarcioglu M, Thuraisingham B (2013) Preventing private information inference attacks on social networks. IEEE Trans Knowl Data Eng 25:1849–1862
79. M. Carroll, A. Van Der Merwe, and P. Kotze, "Secure cloud computing: benefits, risks and controls", In 2011 information security for South Africa 2011, p. 1-9. IEEE
80. Dinh HT, Lee C, Niyato D, Wang P A survey of mobile cloud computing: architecture, applications, and approaches. Wirel Commun Mob Comput 13(18):1587–1611
81. Roman R, Lopez J, Mambo M (2018) Mobile edge computing, fog et al.: a survey and analysis of security threats and challenges. Future Gener Comput Syst 78:680–698
82. Article (2013) Available Online at www.jgrcs.info cloud computing and social networks:a comparison study of. 4(3):51–54
83. Almudawi NA (2016) Cloud computing privacy concerns in social networks cloud computing privacy concerns in social networks. no. August, 2016
84. Zhang Z, Sun R, Zhao C, Wang J, Chang CK et al (2017) CyVOD: a novel trinity multimedia social network scheme. Multimed Tools Appl 76(18):18513–18529
85. Yu S, Wang C, Ren K, Lou W (2010) Achieving secure, scalable, and fine-grained data access control in cloud computing. In: Proceedings of the IEEE INFOCOM
86. Bhamare D, Samaka M, Erbad A, Jain R, Gupta L, Chan HA (2017) Optimal virtual network function placement in multi-cloud service function chaining architecture. Comput Commun 102:1–16
87. Zhang X, Zhang Y, Mo Q, Xia H, Yang Z, Yang M, Wang X, Lunand L, Duan H (2018) An empirical study of web resource manipulation in real-world mobile applications. In: Proceedings of the27th security symposium (Security 18), pp 1183–1198
88. Hashem IAT, Yaqoob I, Anuar NB, Mokhtar S, Gani A, Khan SU (2015) The rise of "big data" oncloud computing: review and open research issues. Inf Syst 47:98–115
89. Deka GC, Das PK (2018) Application of virtualization technology in IaaS cloud deployment model. In: Design and use of virtualization technology in cloud computing. IGI Global, pp 29–99
90. Coppolino L, D'Antonio S, Mazzeo G, Romano L (2017) Cloud security: emerging threats and current solutions. Comput Electr Eng 59:126–140
91. Vlajic N, Chowdhury M, Litoiu M (2019) IP Spoofng in and out of the public cloud: from policy topractice. Computers 8(4):81
92. Gupta BB, Joshi RC, Misra M (2012) ANN based scheme to predict number of zombies in a DDoS attack. IJ Network Sec 14(2):61–70
93. Sumitra B, Pethuru C, Misbahuddin M (2014) A survey of cloud authentication attacks and solution approaches. Int J Innov Res Comput Commun Eng 2(10):6245–6253
94. Kumar PR, Raj PH, Jelciana P (2018) Exploring data security issues and solutions in cloud computing. Proc Comput Sci 125:691–697
95. Gupta S, Gupta BB (2018) XSS-secure as a service for the platforms of online social network-based multimedia web applications in cloud. Multimed Tools Appl 77(4):4829–4861
96. Singh S, Jeong Y-S, Park JH (2016) A survey on cloud computing security: issues, threats, and solutions. J Netw Comput Appl 75:200–222
97. Sumitra B, Pethuru C, Misbahuddin M (2014) A survey of cloud authentication attacks and solution approaches. Int J Innov Res Comput Commun Eng 2(10):6245–6253
98. Zhang Y, Chen X, Li J, Wong DS, Li H, You I (2017) Ensuring attribute privacy protection and fast decryption for outsourced data security in mobile cloud computing. Inf Sci 379:42–61

99. Ahmed M, Litchfeld AT (2018) Taxonomy for identification of security issues in cloud computing environments. J Comput Inf Syst 58(1):79–88
100. Basu S et al (2018) Cloud computing security challenges and solutions—a survey. In: Proceedings of the IEEE 8th annual on computing and communication workshop and conference (CCWC), pp 347–356
101. Indu I, Anand PR, Bhaskar V (2018) Identity and access management in cloud environment: mechanisms and challenges. Eng Sci Technol Int J 21(4):574–588
102. Cai F, Zhu N, He J, Mu P, Li W, Yu Y (2018) Survey of access control models and technologies for cloud computing. Clust Comput 22(S3):6111–6122
103. Khalil I, Khreishah A, Azeem M (2014) Consolidated identity management system for secure mobile cloud computing. Comput Netw 65:99–110
104. Wu M, Moon YB (2017) Taxonomy of cross-domain attacks on cyber manufacturing system. Proc Comput Sci 114:367–374
105. Butun I, Erol-Kantarci M, Kantarci B, Song H (2016) Cloud-centric multi-level authentication as a service for secure public safety device networks. IEEE Commun Mag 54(4):47–53
106. Mohit P, Biswas G (2017) Confidentiality and storage of data in cloud environment. In: Proceedings of the 5th international conference on frontiers in intelligent computing: theory and applications. Springer, Berlin, pp 289–295
107. Islam MA, Vrbsky SV (2017) Transaction management with tree-based consistency in cloud databases. Int J Cloud Comput 6(1):58–78
108. Meinig M, Sukmana MI, Torkura KA, Meinel CJPCS (2019) Holistic strategy-based threat model for organizations. Proc Comput Sci 151:100–107
109. Shin S et al (2014) Rosemary: a robust, secure, and high-performance network operating system. In: Proceedings of the 2014 ACM SIGSAC conference on computer and communications security, New York, pp 78–89
110. Islam MA, Vrbsky SV (2017) Transaction management with tree-based consistency in cloud databases. Int J Cloud Comput 6(1):58–78
111. Somani G, Gaur MS, Sanghi D, Conti M, Buyya R (2017) DDoS attacks in cloud computing: issues, taxonomy, and future directions. Comput Commun 107:30–48
112. Tan Y, Wu F, Wu Q, Liao XJTJOS (2019) Resource stealing: a resource multiplexing method formix workloads in cloud system. J Supercomput 75(1):33–49
113. Iqbal S, Kiah ML, Dhaghighi B, Hussain M, Khan S, Khan MK, Choo KKR (2016) On cloud security attacks: a taxonomy and intrusion detection and prevention as a service. J Netw Comput Appl 74:98–120
114. Yamin MM, Katt B, Sattar K, Ahmad MB (2019) Implementation of insider threat detection system using honeypot-based sensors and threat analytics. In: Future of information and communication conference. Springer, Berlin, pp 801–829
115. Subramanian N, Jeyaraj AJC, Engineering E (2018) Recent security challenges in cloud computing. Comput Electr Eng 71:28–42
116. Meinig M, Sukmana MI, Torkura KA, Meinel CJPCS (2019) Holistic strategy-based threat modelfor organizations. Proc Comput Sci 151:100–107
117. Akshaya MS, Padmavathi G (2019) Taxonomy of security attacks and risk assessment of cloud computing. In: Peter J, Alavi A, Javadi B (eds) Advances in big data and cloud computing. Advances inintelligent systems and computing, vol 750. Springer, Singapore
118. Yamin MM, Katt B, Sattar K, Ahmad MB (2019) Implementation of insider threat detection system using honeypot-based sensors and threat analytics. In: Future of information and communication conference. Springer, Berlin, pp 801–829
119. Bhagwani H, Negi R, Dutta AK, Handa A, Kumar N, Shukla SK (2019) Automated classification of web-application attacks for intrusion detection. In: Lecture notes in computer science, pp 123–141
120. Gumaei A, Sammouda R, Al-Salman AMS, Alsanad A (2019) Anti-spoofng cloud-based multispectral biometric identification system for enterprise security and privacy-preservation. J Parallel Distrib Comput 124:27–40

121. Mishra A, Gupta N, Gupta BB (2021) Defense mechanisms against DDoS attack based on entropy in SDN-cloud using POX controller. Telecommun Syst:1–16
122. Stergiou CL, Psannis KE et al (2020) IoT-based big data secure management in the Fog over a 6G wireless network. IEEE Internet Things J
123. Alsmirat MA, Al-Alem F, Al-Ayyoub M, Jararweh Y et al (2019) Impact of digital fingerprint image quality on the fingerprint recognition accuracy. Multimed Tools Appl 78(3):3649–3688
124. Dahiya A, Gupta BB (2021) A reputation score policy and Bayesian game theory based incentivized mechanism for DDoS attacks mitigation and cyber defense. Future Gener Comput Syst 117:193–204
125. Bhushan K, Gupta BB (2019) Distributed denial of service (DDoS) attack mitigation in software defined network (SDN)-based cloud computing environment. J Ambient Intell Humaniz Comput 10(5):1985–1997
126. Yu C, Li J, Li X, Ren X et al (2018) Four-image encryption scheme based on quaternion Fresnel transform, chaos and computer-generated hologram. Multimed Tools Appl 77(4):4585–4608
127. Hossain MS, Muhammad G, Abdul W, Song et al (2018) Cloud-assisted secure video transmission and sharing framework for smart cities. Futur Gener Comput Syst 83:596–606
128. Gou Z, Yamaguchi S et al (2017) Analysis of various security issues and challenges in cloud computing environment: a survey. In: Identity theft: breakthroughs in research and practice. IGI Global, pp 221–247
129. Al-Qerem A, Alauthman M, Almomani A et al (2020) IoT transaction processing through cooperative concurrency control on fog–cloud computing environment. Soft Comput 24(8):5695–5711
130. Pasupuleti SK (2019) Privacy-preserving public auditing and data dynamics for secure cloud storage based on exact regenerated code. Int J Cloud Appl Comput (IJCAC) 9(4):1–20
131. Kaushik S, Gandhi C (2019) Ensure hierarchal identity-based data security in cloud environment. Int J Cloud Appl Comput (IJCAC) 9(4):21–36
132. Olakanmi OO, Dada A (2019) An efficient privacy-preserving approach for secure verifiable outsourced computing on untrusted platforms. Int J Cloud Appl Comput (IJCAC) 9(2):79–98
133. Cvitić I, Peraković D, Periša M, Husnjak S (2019) An overview of distributed denial of service traffic detection approaches. Promet – Traffic Transport 31(4):453–464
134. Cvitić I, Peraković D, Periša M, Botica M (2019) Novel approach for detection of IoT generated DDoS traffic. Wireless Network
135. Cvitic I, Peraković D, Periša M, Jurcut AD (2021) Methodology for detecting cyber intrusions in e-learning systems during COVID-19 (in press). Mobile Networks Appl
136. Perakovic D, Perisa M, Cvitic I, Husnjak S (2017) Artificial neuron network implementation in detection and classification of DDoS traffic. Telfor J 9(1):26–31
137. Dahiya A, Gupta BB (2021) How IoT is Making DDoS Attacks More Dangerous?, Insights2Techinfo, pp.1
138. Gupta BB, Joshi RC, Misra M, Jain A, Juyal S, Prabhakar R, & Singh AK (2011, April). Predicting number of zombies in a DDoS attack using ANN based scheme. In International Conference on Advances in Information Technology and Mobile Communication (pp. 117-122). Springer, Berlin, Heidelberg

Chapter 15
Monitoring the Energy Consumption of FDM Device Based on the Variation of Operating Parameters: A Study

Jozef Husár and Jakub Kaščak

1 Introduction

There must be a 3D digital model at the beginning of each 3D printing process. This model is created in various 3D programs. In the industry, there are mainly various 3D CAD programs for business, but also for ordinary end users, there are cheaper and easier to use programs, or programs which can directly use scanned data from 3D scanners. The next kind of used programs are so-called slicers in which the models "cut" into individual layers and is thus ready to be sent to a 3D printer. Subsequently, the printer applies material to the individual layers depending on the shape of the model and the printing process. Of course, there are several types of 3D printing technologies that use different materials in different ways to create the final object. Today, there are different types of plastic, metal, ceramics, and sand powder (similar to artificial sandstone) that are available as materials. Research is currently also focused on the use of biomaterials and various types of materials suitable for consumption [1]. However, it should be noted that in the cheaper segment of 3D printers intended for the general public, the use of materials for 3D printing is still quite limited to plastic. Plastic is currently the most widely used material – mainly Acrylonitrile butadiene styrene (ABS) or Polylactic acid (PLA), although the number of alternative materials such as Nylon, is growing. As mentioned in the abstract,

J. Husár (✉)
Faculty of Manufacturing Technologies with aseat in Prešov, Department of Industrial Engineering and Informatics, Technical University of Košice, Prešov, Slovak Republic
e-mail: jozef.husar@tuke.sk

J. Kaščak
Faculty of Manufacturing Technologies with aseat in Prešov, Department of Computer Aided Manufacturing Technologies, Technical University of Košice, Prešov, Slovak Republic
e-mail: jakub.kascak@tuke.sk

L. Knapcikova et al. (eds.), *Sustainable Management of Manufacturing Systems in Industry 4.0*, EAI/Springer Innovations in Communication and Computing, https://doi.org/10.1007/978-3-030-90462-3_15

we focus on FDM printing of thin-walled objects and monitoring of the energy consumption of 3D printing process. First, it is necessary to define a few basic components and parameters of 3D printers [2].

Each FDM device consists of several key components that are characteristic of its construction. Of course, there are types of devices with multiple functions and their components are more diverse. This chapter characterizes all the basic elements which the FDM device consists of, their function within the structure, and the basic principle of operation [3].

3D Printer Frame

At present, there are different types of 3D printers available in different price categories. These devices are constructed of inexpensive and affordable components, such as threaded rods in combination with various plastic parts, or parts of different materials. However, these types of constructions do not provide sufficient accuracy during assembly or sufficient rigidity during movement. Mid-range printers are made of various types of sheet metal in combination with a plastic design cover. These are the most ideal for the average user in terms of availability and price ratio. Industrial machines are made of a steel base frame or aluminum profiles supplemented by a plastic or sheet metal cover. These parts guarantee sufficient rigidity and resistance to the high temperatures that often occur inside the working space of these devices. Fig. 15.1 shows an example of the simplest constructions, down to Stratasys professional equipment [4].

1.1 Stepper Motors

The constructions of FDM devices are largely inspired by the movements and design of other computer numerical control (CNC) devices. Among these devices are milling machines, laser engravers, industrial manipulators, and robots. The most

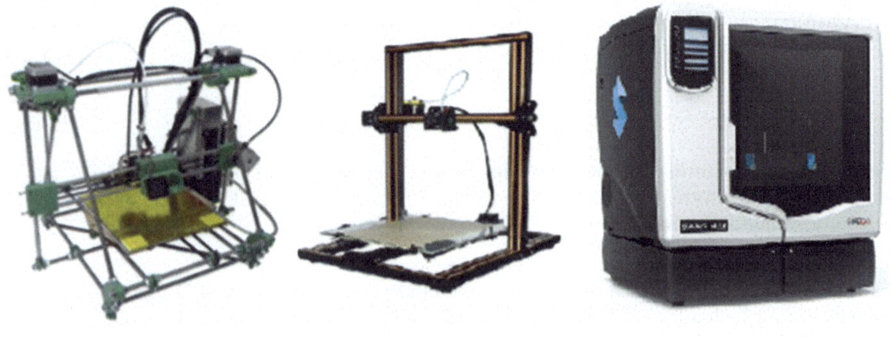

Fig. 15.1 Selected types of constructions (from the left side: frame made of threaded rods (first Reprap printers), frame made of aluminum profiles (commercially available printers), Stratasys uPrint professional equipment)

common type of construction for FDM devices is currently a system operating in the cartesian coordinate system. Structures of this type move each axis separately. In general, the x-axis moves relative to the device, from left to right, the y-axis forwards and backwards, and the z-axis from bottom to top. Stepper motors are used for movement and lead screws or timing belts are used as transformation mechanisms. Commercial hobby equipment uses stepless motors for movement in all three axes. The most accurate and expensive models also use stepper motors with feedback, but this solution requires a different type of electronics and more complex control software. Stepper motors with a step length of at least 1.8° are considered sufficient for simple FDM-type devices. These are controlled by means of controls and a mainboard. The basic one allows the control of stepper motors with a maximum current of 2A, and their static torque is in the range of 0.5–2.5 Nm. This definition is characteristic of standard NEMA 17 engines, which are among the most widely used in the field of commercially available FDM equipments. The smooth flow of stepper motors is ensured by base plates, which can decompose their movement into so-called micro-steps, most often in a ratio of 1/16 or 1/32. At present, there are commercial and available devices with base plates capable of handling up to 1/256 microsteps [5].

The new concepts also implement linear stepper motors in their constructions. These are characterized by their higher dynamics, speed of movement, and accuracy compared to commonly used rotary stepper motors (Fig. 15.2).

Within the actuators, several variants are available which are suitable for use with FDM devices. These include permanent magnet motors, variable reluctance, hybrid, unipolar, and bipolar motors. Motors with variable reluctance are characterized by a toothed rotor which is mounted between the stator poles, thus achieving rotation. Similar to permanent magnet motors, each of the windings is powered separately, so the resulting polarity ensures rotational movement [6]. Hybrid motors combine individual motor elements with a permanent magnet and variable reluctance. The rotor of such a motor is divided into two parts, these have the opposite

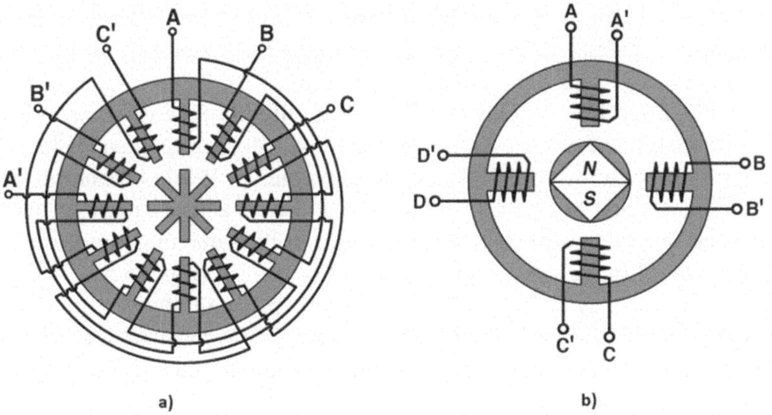

Fig. 15.2 Variable magnet motor (**a**) [6], variable reluctance motor (**b**) [5]

Fig. 15.3 Hybrid stepper motor [5]

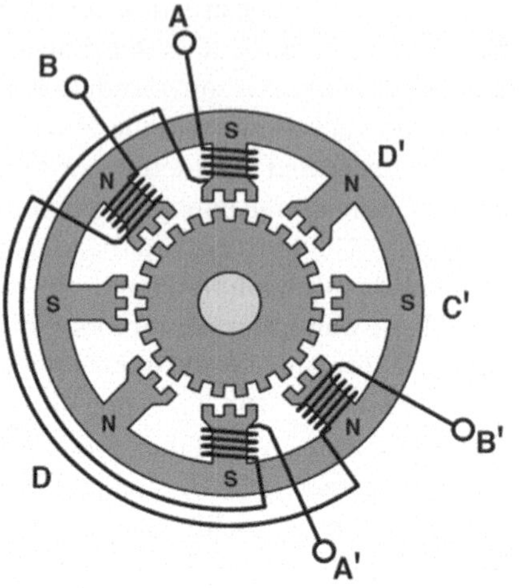

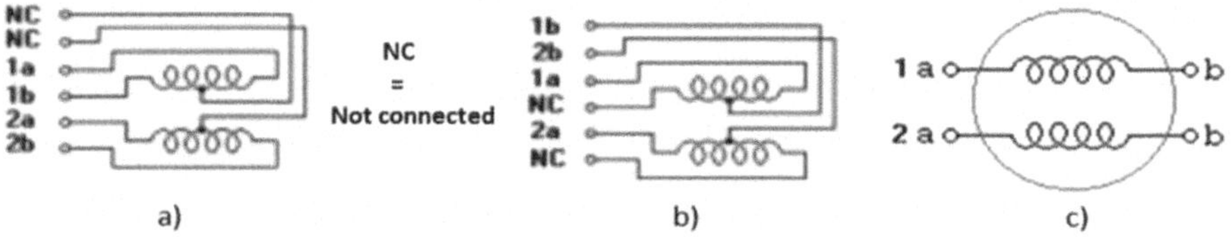

Fig. 15.4 Wiring diagram of unipolar SM as bipolar, in order to achieve higher torque, c) connection of coils of bipolar stepper motor

polarity and gearing, fitted in the same way as in reluctance motors. Reluctance and hybrid motors are able to achieve higher accuracy compared to permanent magnet motors. This lack of accuracy is compensated for in permanent magnet and hybrid motors by the possibility of using the so-called micro-step algorithms [5] (Fig. 15.3).

Unipolar and bipolar motors are other possible variations of hybrid motors and permanent magnet motors. Bipolar motors are very similar in physical equipment to unipolar ones, except that they do not have an outlet that extends from the center of the vessels between its ends, as can be seen in Fig. 15.4.

Due to this difference, bipolar motors need a different type of control than unipolar. With this type of electric motor, it is necessary to ensure changes in the flow of the supply through the coils by changing the polarity. While bipolar motors use the entire coil, not just one half, they have the ability to achieve more torque. With unipolar control, it is possible to achieve a torque of the same magnitude only in the four-stroke control mode, i.e., powering the entire coil [7] (Fig. 15.5).

Manual control is available on the device to control all stepper motors. If the device is compatible with a PC, software control of all stepper motors is possible, including the motor that controls the extrusion of the material [8] (Fig. 15.6).

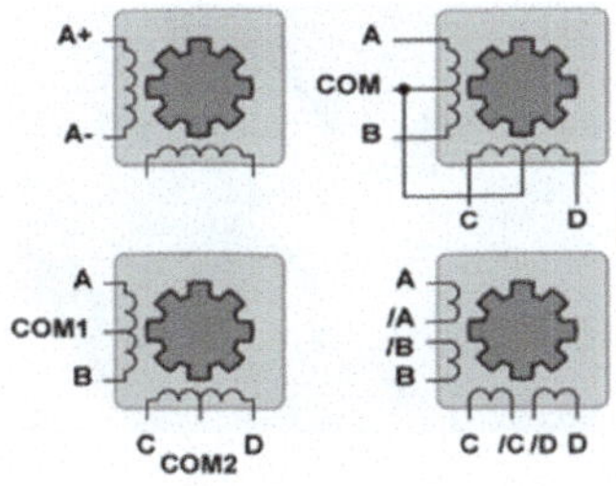

Number of conductors	Possible ways of excitation of phase windings
4	Bipolar
5	Unipolar
6	Unipolar or bipolar (winding connected in series)
8	Unipolar (winding connected in series), bipolar (winding connected in series or in parallel)

Fig. 15.5 The most common ways of connecting phase windings of stepper motors

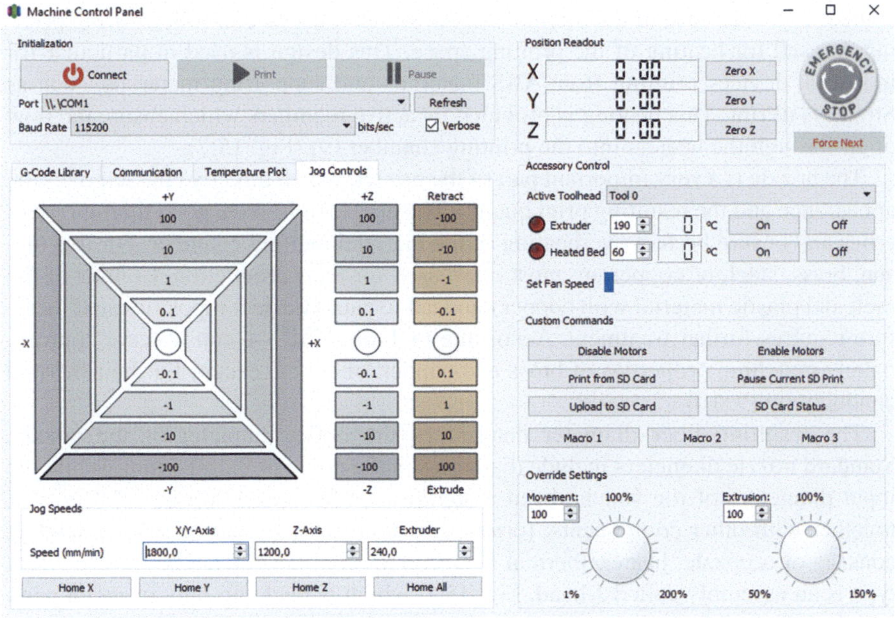

Fig. 15.6 Simplify3D software interface designed to control the device's stepper motors

1.2 Printhead and Heating System

During the printing process, we need to maintain control, depending on the design of the device, of two to three thermal circuits. The first circuit is the heating of the print head. It can work with temperatures up to 350 °C, 12 and 24V power supplies are used for its powering. The second circuit is the heating of the printing bed, or so-called heatbed. This ensures better adhesion of the printed object to the substrate. It thus enables the printing of different materials or printing from several types of materials in one printing process, by preventing the print from peeling off the substrate and shrinking during cooling. For this purpose, Polychlorinated biphenyl (PCB) or aluminum plates are used in combination with PCBs, which serve as

Fig. 15.7 Types of heated plate constructions used in FDM-type devices

heating elements. Examples seen in Fig. 15.12 are heatbeds, powered by a 24V source. The third case of the thermal circuit, which must be controlled during printing, is used for heating of the printing space. This design is used in particular for industrial devices printing from ABS materials and supporting materials, such as Stratasys uPrint. This heating is provided by active members which ensure the flow of air through the heaters into the printing chamber [9] (Fig. 15.7).

The nozzle is a very important part of the printer, which directly characterizes the appearance and the resulting print quality. The nozzle must have good thermal properties and ensure perfect overheating of the material melting chamber. For this reason, brass, steel, or copper are most often used for their production. Contact of the melt and plastic material with copper can lead to catalytic decomposition and therefore requires further treatment. According to Table 15.1, we can observe approximately matching properties of brass and copper, of course except for their thermal conductivity [9, 10].

The resulting fiber diameter determines the outlet diameter of the nozzle. Standard nozzle diameters include diameters in the range of 0.2–0.5 mm. While the input diameters of the standardized material are 1.75 and 3.00 mm. The nozzle, together with other components, forms part of an FDM device called a hotend. It consists of a nozzle, heater, thermal bridge, and passive cooler. The most modern type is an assembly called J-head, Fig. 15.8, which is made up only of metal parts and thus allows higher temperatures to be achieved during the melting of the material [11].

The heated block is heated to temperature by means of a thermal resistor and transfers heat to the nozzle by conduction. At its center, it melts the fiber to the desired temperature. The melt is then pushed through the nozzle onto the heatbed by the force of the extruder. The aim of this activity is to melt the fiber in the shortest possible section, which is aided by thermal bridging and passive cooling. The purpose of the thermal bridge is to prevent the possible spread of heat toward the radiator. This dissipates residual heat to the surroundings and thus prevents the material from melting at a greater distance from the nozzle. In the event of failure and melting of the fiber above the limit temperature, still in the area of the cooler, the plasticity of the material will cause its inaccurate dosing. This will affect the resulting print quality, resp. on printing errors [9].

Table 15.1 Thermal properties of materials used for nozzles of FDM equipment

	Thermal conductivity, λ [$W.m^{-1}.K^{-1}$]	Density, ρ[$kg.m^{-3}$]	Melting point, Tm [°C]	Specific heat capacity, C [$J.kg^{-1}.K^{-1}$]
Copper	386	8900	1064	390
Aluminium	237	2700	660	920
Brass	120	8400	850–920	385
Steel	50	7850	1539	460

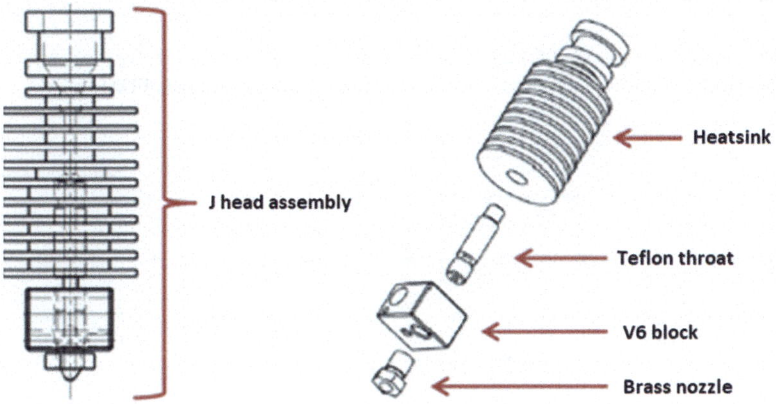

Fig. 15.8 Example of J-head hot end, with a description of its parts

The principle of the fiber extrusion function system is to induce one-sided pressure on the fiber, exerted by an extruder grooved wheel on the material. These grooves cause cutting into the material and its subsequent displacement by means of the pressure exerted from the compression spring. Due to the movement of stepper motors shaft, on which the splined wheel is located, this process is software controlled and can be modified during print. The construction of the extruder type is used in two versions, the first is the version with the so-called direct extruder, which is located directly on the printhead, i.e., on the x-axis travel. The second option is a bowden design, i.e., the extruder, is not located directly on the hotend, but outside it. The connection between the hotend and the extruder is ensured by means of a teflon tube, which can be seen in Fig. 15.9. This design is used especially for constructions that do not have sufficiently strong frames, which reduces the weight that the device must handle [6].

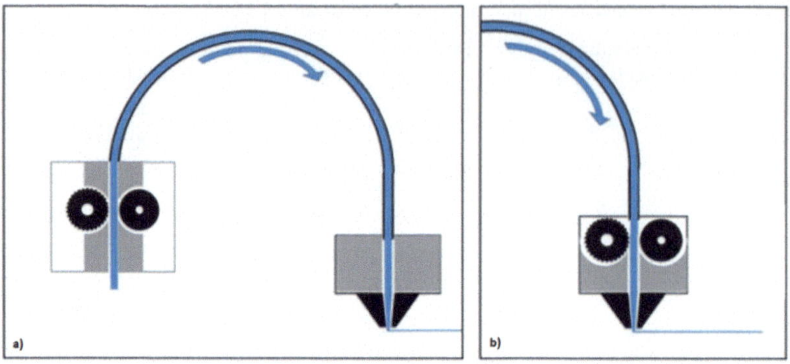

Fig. 15.9 Types of printheads FDM devices (**a**) Bowden printing system with extruder outside the printhead, (**b**) Direct printing system with extruder directly on the printhea FDM (**a**)

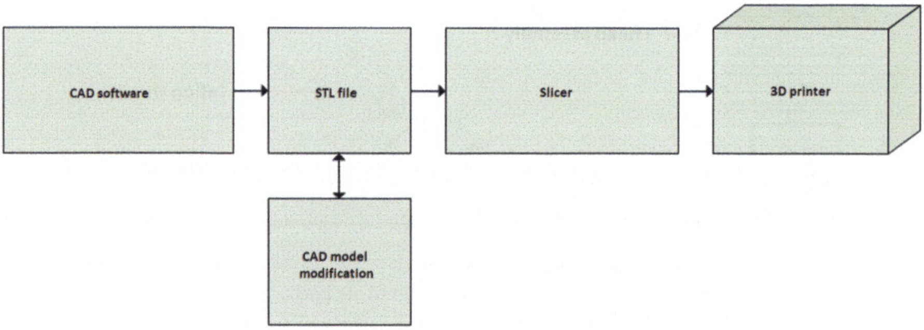

Fig. 15.10 Communication scheme between software and FDM device

1.3 Control Units and Power Supplies

The first step before printing is to prepare the model. During this process, the object is exported from the CAD program in STL format. This object is cut by software into individual layers perpendicular to the vertical z-axis, and in each of the layers, a system of movements is generated which copy the geometry of the printed object in a given section plane. The motion system is generated in the form of a G-code, which is then sent by a software to the printer control unit. The most widespread freely available programs ensuring communication between the software and the control unit include Cura, Repetier host, Simplify3D, and more. These programs ensure the sending of the G-code to the printer control unit, which then processes it using the implemented firmware. The scheme of mutual communication between the individual components can be seen in Fig. 15.10 [12].

The firmware contains all the necessary data on the settings and specifications of the FDM device used. These settings include setting of precise motor stepping,

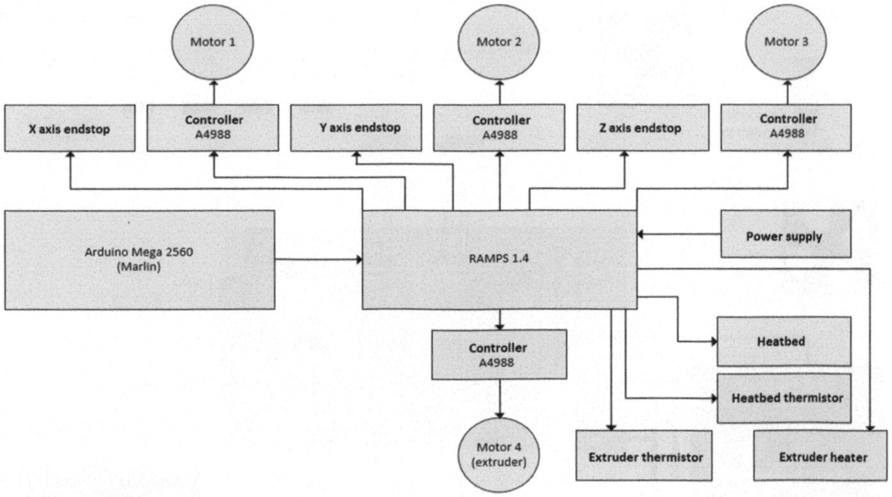

Fig. 15.11 Wiring diagram of Arduino Mega 2560 with RAMPS

coordinate conversions, safety temperature limits, and maximum travel path lengths. There are several types of firmware available, such as Marlin, Sprinter, or Teacup. As far as the architecture of mainboard is concerned, several variations appear on the market, but the most frequently used combination is the Arduino mega 2560 series and the RAMPS. Other control units used include Rambo, Melzi, and Rumba. The connection between the PC and the control board is most often secured using a USB cable. At present, communication via Wi-Fi network is already used. The wiring diagram of the individual elements of the FDM device using the Arduino Mega 2560 and its RAMPS extension is described in Fig. 15.11.

A real view of the individual components of the FDM device and their connection with one nozzle, heatbed, and control LCD panel can be seen in Fig. 15.12. Important components include limit switches, which like CNC machines, allow setting the exact position of the end point, which serves as a reference point to determine the origin of the coordinate system. These switches can be mechanical or optical. An indisputable advantage is the optical sensors, which allow sensing by the free passage of the aperture between the optical sensor and the diode. This sensor is advantageous especially at higher speeds of approaches to zero positions, where due to inertial forces, an impact and its destruction can occur when using a mechanical sensor. The lifespan of both types of sensors is limited by a certain number of cycles.

The devices analyzed in this chapter are characteristic of FDM technology, but it is not a complete list of components. In connection with the complex architecture of FDM devices, which contain a series of different components involved to some extent in its operation. Other FDM devices include: connecting cables, screw nuts, LCD display, 12 or 24V power supply, fans, components for transformation, and motion transmission [13] (Fig. 15.13).

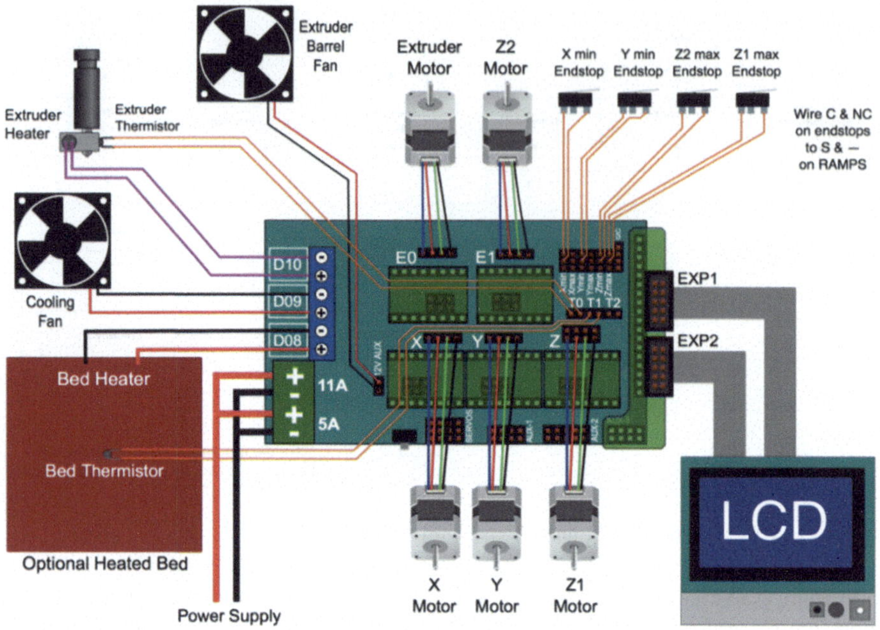

Fig. 15.12 Wiring diagram of the arduino 2560 control unit and individual components of the FDM device [13]

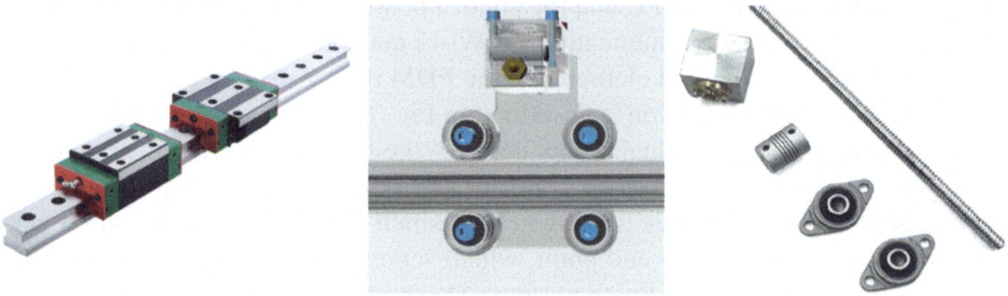

Fig. 15.13 Demonstration of transformation mechanisms used by FDM technology

We can evaluate the energy requirements of the FDM devices by ourselves, for example, according to the Creality CR-10 max device, the individual parts of which were used to build the basis of our design. The Creality CR-10 max with its above-standard print volume of 0.095 m^3 consumes a relatively small amount of energy. Its maximum value is at the limit of 170W and the average value is at the level of 105W during the printing process. In standby mode, i.e., when the device is switched on, the average value is at the level of 15.7W, which corresponds to a laptop in sleep mode. As for the performance of individual components, the biggest consumer is the heatbed with a size of 300 x 300 mm, which requires up to 129.5W to heat up to 90 °C. This temperature is used when printing from ABS material and is usually the

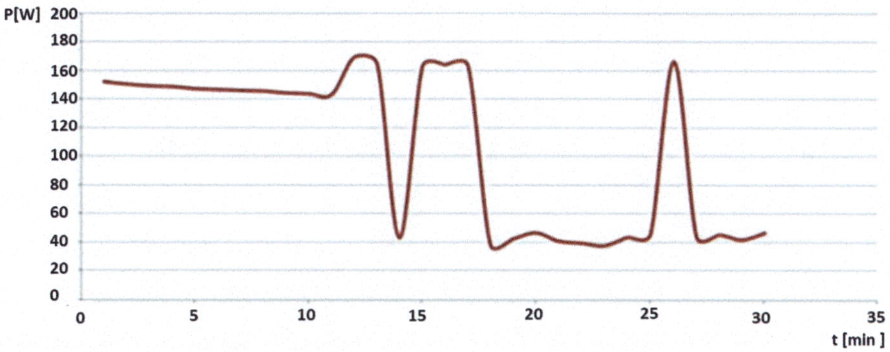

Fig. 15.14 Power fluctuations during normal operation of the FDM device

maximum possible temperature that commercial FDM devices can reach during their operation. Another consumer of energy is the hotend, i.e., the heating part of the print head requiring 28.3 W and subsequently stepper motors in individual axes. For a motor in the x-axis = 0.3W, y-axis = 0.4W and 2W for the lifting z-axis, where two motors operate simultaneously [13].

Figure 15.14 shows the variation in printing power of the 40 x 60 x 20 mm model. From the resulting graph, there is a clear increase in the consumption of electricity at the very beginning of the press. This increase is due to the extreme consumption that occurs when the platen is heated.

1.4 Production Speed

The main factor that significantly limits the use of FDM devices is their production speed. Figure 15.15 shows a model of a stress test used to calibrate the operating parameters of FDM devices. Its maximum dimensions are $x = 75$ mm, $y = 70$ mm, $z = 92$mm. Using a standard 0.4mm diameter nozzle and 20% fill, the time required to print this model is up to 3 h and 23 min. The two main factors influencing the printing speed of the models are considered to be the maximum speed that used stepper motors can achieve and the maximum speed of material extrusion. The speed at which the extruder can extrude molten thermoplastic material is significantly slower than the speed of movement of stepper motors. Since the standard Nema 17 stepper motors can reach speeds between 40 and 150 mm/s during printing, it can be argued that the parameter that limits the printing speed of the models is the extrusion speed. Of course, there are number of parameters and options that allow you to manipulate the time required for printing. Some of them are: replacement of the nozzle with a nozzle of larger diameter, use of a lower percentage of filling, and reduction of circumferential perimeters. However, all these parameters, as well as the increase of the movement speed in each axis, have a great influence on the final quality of prints [14, 15].

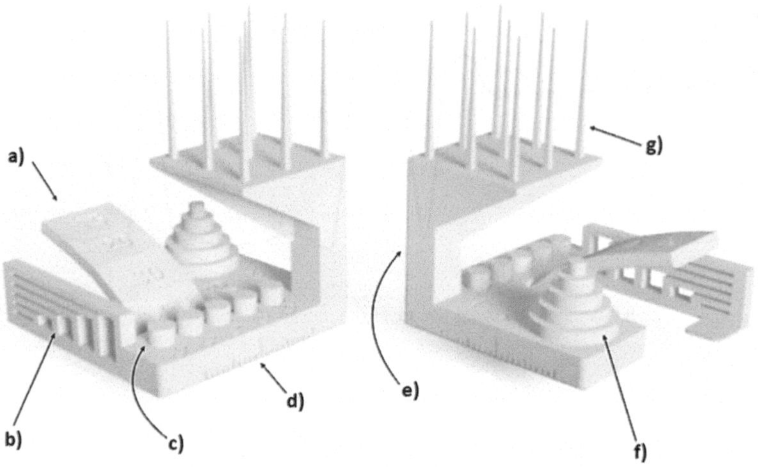

Fig. 15.15 Example of a comparative test of FDM devices used in their calibration (based on the shown areas, (**a**) overhangs are checked, (**b**) bridges without support, (**c**) dimensions of negative elements, (**d**) stiffness in x and y axes, (**e**) z-axis alignment, (**f**) dimensional accuracy, (**g**) extrusion and retraction settings

1.5 Production Quality

The design of the FDM device, which is adapted to create the model from the heat-bed upward, brings several complications. The biggest of them can be considered as the very fact that the device can not create a model that has overhangs greater than 45°. Overhang with an angle greater than 45° requires the use of supports that prevent deformation of the model or the occurrence of a possible failure. The creation of overhangs approaching the limit of 45°, but also the use of the supports themselves, results in deformation of the surface of the models in almost all types of FDM devices. In addition, the mere use of supports, whether of the same material as the model itself or of another type of removable material, rapidly increases the time required to produce the model [13] (Fig. 15.16).

The dimensional and shape accuracy of the models is directly related to several parameters and the construction elements themselves. While the use of support respectively subsequent disposal can cause damage to the model during production using conventional types of FDM equipment. It is possible to partially reduce this type of damage by setting the device control software correctly. However, some parameters influencing the final quality of the printout are directly related to its design. For example, in the planar printing of models by means of conventional devices, even at the lowest possible layer height, an often undesirable surface roughness occurs. This roughness is equal to the height of the layer on the surface of the whole model. The height of this layer ranges from 0.05–1.2 mm [13].

The quality of the model depends mainly on the height of the layer and the total printing time. The smaller the layer height, the less noticeable scratches on the

Fig. 15.16 Demonstration of support structures used by FDM technology

Fig. 15.17 Examples of product quality depending on layer height and printing time [17]

object. All this affects the head of the printer, which has to move along a different perimeter. This is directly related to the printing time, which is longer [16] (Fig. 15.17).

2 Methodology

To verify the theoretical part, we chose to manage models into objects called litophanes. It is basically processing photos into 3D models. Their specific feature is their partial transparency when illuminated from one of the sides. This type of model was chosen for the purpose of a simple visual inspection of the accuracy of the layering of the material, which will also allow the detection of possible defects in the layering in an equally simple manner [18]. Our goal is to focus on the values of electricity consumption of FDM additive production equipment at different layer heights. The purpose of this study is to collect the data needed to optimize this type of production from an economic and production time point of view. The study serves as a basis for the gradual streamlining of FDM production and is an initial step to optimize the entire model production process.

2.1 Hardware and Software

Hardware

The printer used for printing is Creality CR- 10 max with the following parameters:

- Technology: FDM
- Filameter diameter: 1.75 mm
- Compatible materials: PLA, PETG, TPU, wood, other exotics
- Accuracy: x / y-axes +/- 12 microns, z-axis 1 micron
- Layer height: 100–400 microns
- Feeder system: Bowden
- Extruder type: Single
- Nozzle size: 0.4 mm, 0.8 mm
- Max. extruder temperature: 250 °C
- Max. heated bed temperature: 100 °C
- Build volume: 450 × 450 × 470 mm
- Printer dimensions: 735 × 735 × 305 mm

For the purpose of this test, the selected PETG material was characterized by the following values: diameter 1.75 mm with tolerance 0.05 mm, weight (full spool) 1 kg, flexibility, high temperature resistance, highest suitable printing temperature up to 260 °C, substrate temperature 80 °C, recommended print speed from 40 mm/s to 60 mm/s, lower speeds are used for lithophane printing [19].

We used the Eve Evergy smart socket to measure the energy consumed. This socket is characterized by high safety and ease of operation. Eve Energy controls lights and appliances with a simple tap or with an application that supports Apple Homekit. The advantage is that it makes it possible to determine exactly how much energy the individual devices connected to it use. Technical parameters [20]:

- Smart plug & power meter
- Bluetooth Low Energy wireless connection
- Works with Apple HomeKit
- for European sockets type E and F and appliances type C and F
- AC input 100–240 V, 50/60Hz
- AC 100–240 V output
- max. 11A/2500W
- switches appliances on / off, monitors electricity consumption (Fig. 15.18)

Fig. 15.18 Creality CR10 max with PETG filament

Software Equipment
Monitoring these processes is based on 4 types of software. The first is standard CAD software that creates a 3D model to be printed. In our case, it was PTC Creo parametric. The second software is an application accessible on the internet, freeware in which a 2D image is transferred to a 3D model. The third software is a slicer in which a G-code for a printer is created from a 3D model. The latest application is Eve for Homekit which is used to measure power consumption.

2.2 *Work Methodology*

Our first step was to build the desired pattern that we wanted to print. For our needs, it was the logo and name of the Technical University in Košice (Fig. 15.19).

Subsequently, we transferred this photo to a web application in which we entered the basic requirements for a 3D model. These were specified as follows:

Image settings:

- Positive image
- Mirror Umage Off
- Flip Image Off
- Mirror Repeat Off
- Flip Repeat Off
- Model settings:
- Maximum size: 50mm
- Thinkness: 3 mm
- Border: 3mm

Fig. 15.19 Photo from the process of printing a test sample

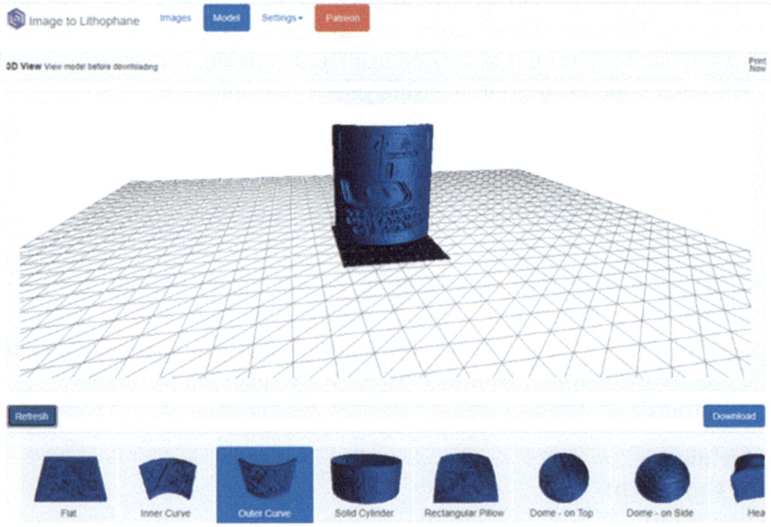

Fig. 15.20 Web application and 3D model for lithophane [18]

- Thinnest Layer: 0.8 mm
- Vectors Per Pixel: 4
- Curve: 0 (Fig. 15.20)

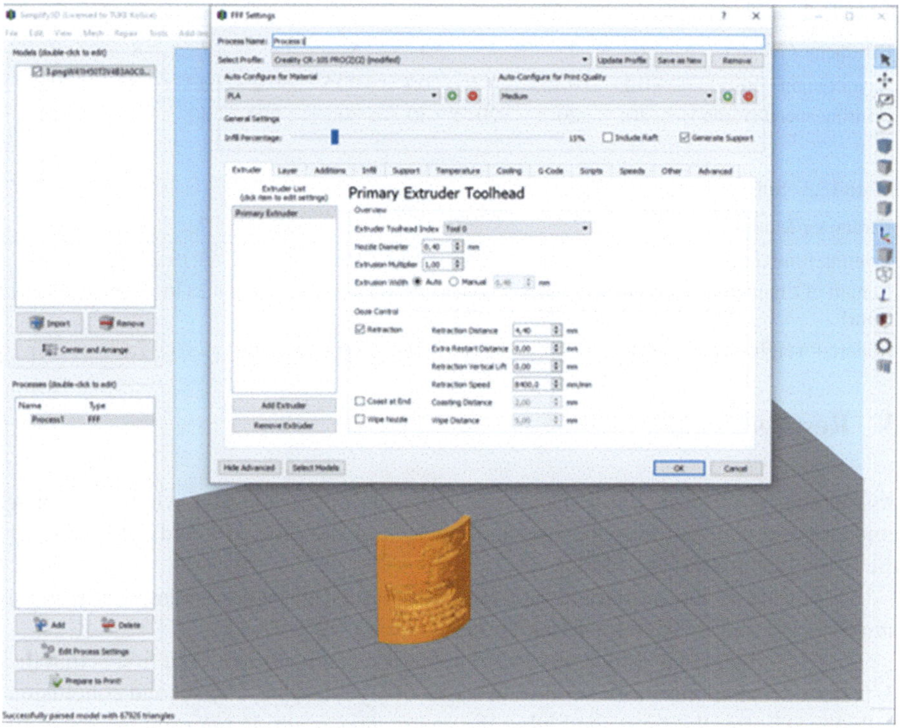

Fig. 15.21 Slicer simplify 3D basic settings

After generating a 3D model in STL format, we imported the model into a slicer and created 9 models in which we changed the parameters of the basic input parameters such as layer height and print speed (Fig. 15.21).

The following parameters were chosen as the basic invariant parameters for all 9 model models:

- Infil: 100%
- Nozzle diameter: 0.4mm
- Skirt: 5
- Board temperature: 70 °C
- Nozzle temperature: 250 °C
- Material: PETG Plasty Mladeč
- Room temperature 22 °C

The units that were unchanged during printing are listed in Table 15.2.

After generating 9 models, we examined the software-generated printing time, consumption, and weight of the material (Table 15.3).

Table 15.2 Basic parameters of models

Parameter / Model	1	2	3	4	5	6	7	8	9
Layer height (mm)	0,1	0,15	0,2	0,1	0,15	0,2	0,1	0,15	0,2
Printing speed (mm/s)	20	20	20	30	30	30	40	40	40

Table 15.3 Software generated data

Parameter/Model	1	2	3	4	5	6	7	8	9
Printing time (hod)	2:48	1:52	1:25	2:19	1:33	1:10	2:19	1:33	1:10
Length of filament (mm)	2336,9	2339,4	2347,1	2336	2339,4	2347,1	2336	2339,4	2347,1
Material weight (g)	7,03	7,03	7,06	7,03	7,03	7,06	7,03	7,03	7,06

3 Results

As already mentioned, the main idea of this article is to identify what the energy consumption was and how the speed and height of the print affected the quality of the overall printout (Fig. 15.22).

After printing the individual models, we recorded energy consumption using a smart socket and the Eve for Homekit application (Table 15.4).

Based on the actual values, several conclusions can be drawn. The first is that the 3, 6, 9 models had the shortest times in terms of layer height. The second conclusion is that the shorter the time, the higher the energy consumption per minute. The third conclusion is that when starting the printing, a higher consumption of up to 4.9 W/min was created. This was caused by the first heating of the nozzle to 250 °C and the plate to 70 °C. Subsequently, the temperature was kept at the same values, which is confirmed by the figure in Fig. 15.23.

Quality of Prints Created with Different Layer Heights

As can be seen from Fig. 15.24 in this section, we focused on the identification of defects and the print quality with consideration to the speed and height of print layer. As is obvious, the most defects in accuracy of image occurred at a layer height of 0.2 mm. This is confirmed by models 3, 6, and 9. On the contrary, for comparison, starting from Fig. 15.19, we achieved the highest print quality at a layer height of 0.1 mm and a print speed of 20 mm/s. As it is obvious, during printing, we had several defects such as burn-in of model 2, shown in Fig. 15.24. We also had several defects due to the accumulation of material.

Fig. 15.22 Models of the different heights of the layer printed by the lithophane method

Table 15.4 Actual power consumption and printing time

Parameter/model	1	2	3	4	5	6	7	8	9
Total consumption (W/h)	752,5	503,36	412,87	648,04	449,97	344,57	646	447,85	355,31
Real printing time (hod)	3:07	2:07	1:38	2:40	1:49	1:23	2:38	1:48	1:24
Average consumption energy/min (W/min)	4,02	3,96	4,21	4,05	4,12	4,15	4,08	4,14	4,22

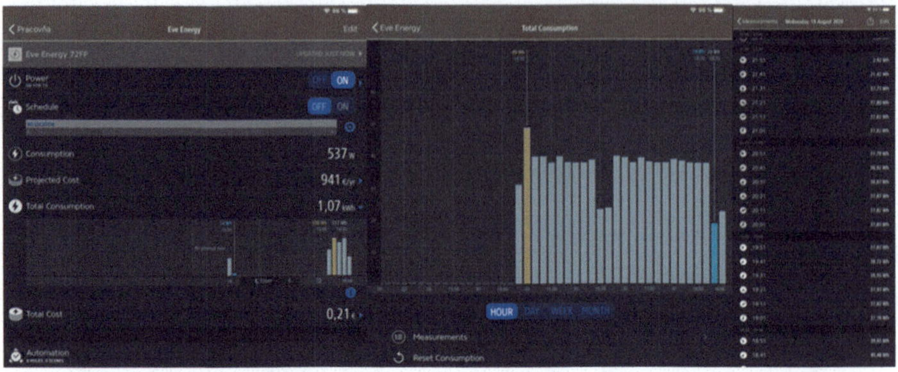

Fig. 15.23 Eve for Homekit environment

4 Conclusion

Nowadays, due to the modern expansion of 3D printing and the perspective of unlimited possibilities, we have tried in this article to show how individual parameters influence 3D printing. Since there are many kinds of materials and also types of printers, we have focused on FDM printing lithophane objects. We used polyethylene terephthalate glycol (PETG) material for the test samples. Lithophane is a 3D print of thin-walled objects based on 2D shapes. We chose one pattern for samples and printed it 9 times, changing the parameters of the layer height in the range of 0.1–0.2 mm and the print speed of 20–40 mm/s at a nozzle temperature of 250 °C. After generating the basic model, we obtained the estimated printing time of individual models in the range of 1:10–2:48 hours. As confirmed by measurements, the actual printing time of the models ranged from 1:23–3:07 hours. this represents an increase in printing time of 10–15%. This can be attributed to the process of calibrating the heatbed and the process of heating the bed and nozzle. Other monitored parameters, which were the total energy consumption per hour and the average energy consumption per minute, showed us that the energy consumption increased with respect to total print length. This means that the less time it takes to print the sample, the higher is the consumption. The last parameter evaluated by us was print quality. Here we can say a simple conclusion that the slower the print and lower layer height, the better the print quality. However, efficiency and quality requirements must also be taken into account. Sometimes the required quality is satisfactory and we can achieve a significant difference in total time and consumption. For comparison, Model 1 and Model 7 are almost the same quality but the savings are 106 W/h and 29 min of printing. Another direction of the authors is the possibility to perform similar measurements when changing the material to PLA and ABS. For PLA, we assume similar time, but the consumption will be lower because the board does not need to be heated to 70 °C and the nozzle temperature will change to 190 °C, which means lower energy consumption.

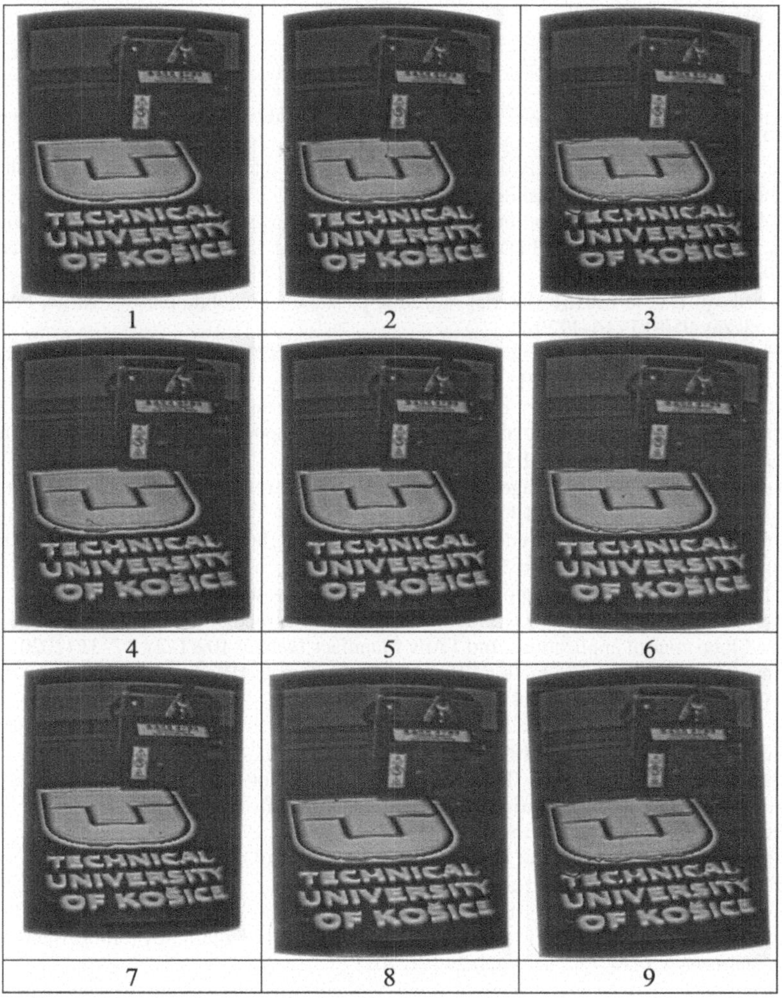

Fig. 15.24 Visual inspection of illuminated lithophane models

Acknowledgements This work was supported by the Slovak Research and Development Agency under the contract No. APVV-19-0590, by the projects VEGA 1/0700/20, KEGA 055TUKE-4/2020 granted by the Ministry of Education, Science, Research and Sport of the Slovak Republic and by the Project of the Structural Funds of the EU, ITMS code: 26220220103.

References

1. Kaščak J et al (2021) Polylactic acid and its cellulose based composite as a significant tool for the production of optimized models modified for additive manufacturing. *Sustainability 13(3)*:1256
2. Židek K et al (2020) Digital twin of experimental smart manufacturing assembly system for industry 4.0 concept. Sustainability 13(9):3658
3. Kaščak J (2020) Research and implementation of an innovative approach to the design solution of device for additive FDM technology, Technical University of Kosice, Disertation work, pp 95
4. Knapčikova L et al (2016) Material recycling of some automobile plastics waste. Przemysl Chemiczny 95(9):1716–1720
5. Stepper motors. http://profirobot.cz/wp-content/uploads/2018/03/Krokov%C3%A9-motory.pdf. Last accessed 24 Feb 2021
6. Buser T. Dissolvable support material used for 3D printing gearbox and Hilbert Cube. http://www.3ders.org/articles/20120128-dissolvable-support-material-used-for-3d-printing-gearbox-and-hilbert-cube.html. Last accessed 24 Feb 2021
7. Tchamna R et al (2016) Management of linear quadratic regulator optimal control with full-vehicle control case study. Intl J Adv Robot Syst 13(5)
8. Flyntt J Benefits of dual extruder printers. 3DINSIDER http://3dinsider.com/dual-extruder-printing/. Last accessed 21 Feb 2021
9. 3D printers. http://tvaroch.sk/blog/3d-tlaciarne/. Last accessed 21 Feb 2021
10. Rivera, M.F.J., Arciniegas, A.J.R: Additive manufacturing methods: techniques, materials, and closed-loop control applications, Intl J Adv Manufact Technol 109(1–2), 17–31 (2020)
11. Kliment M et al (2020) Production Efficiency Evaluation and Products' Quality Improvement using Simulation. Intl J Simulat Model 19(3):470–481
12. Behunova A, Behun M, Knapcikova L (2018) Simulation software support of manufacturing processes in engineering industry. TEM J 7(4):849–856
13. Fused Deposition Modeling; Technical Specifications for FDM. Materialise http://www.materialise.com/en/manufacturing/3d-printingtechnology/fused-deposition-modeling. Last accessed 22 Feb 2021
14. Ahlers D (2018) 3D printing of nonplanar layers for smooth surface generation, Masterthesis, MIN- Fakultät, University of Hamburg, pp 13–19
15. Sher D SpiderBot aiming to extrude perfect double filament webs. 3D Printing Industry. https://3dprintingindustry.com/news/spiderbot-aiming-extrude-perfect-double-filamentwebs-32626/. Last accessed 24 Feb 2021
16. 4 important questions before buying a 3D Printer. http://tobuya3dprinter.com/4-questions-before-buyinghome-3d-printer/. Last accessed 22 Feb 2021
17. Stanko M (2016) The 3D Printer Prusa Mendel Launching, Brno University of Technology, Bachelor work, pp 57
18. Yuan M, Bourell DL (2013) Fundamental issues for additive manufacturing of lithophanes. In: 6th international conference on advanced research in virtual and physical prototyping (VRatP). CRC Press-Taylor & Francis Group, Leira, Portugal, pp 89–93
19. Yuan M, Bourell D (2016) Quality improvement of optically translucent parts manufactured from LS and SL. Rapid Prototyp J 22(1):87–96
20. EVE-ENERGY. https://www.evehome.com/en/eve-energy. Last accessed 22 Feb 2021

Index

L. Knapcikova et al. (eds.), *Sustainable Management of Manufacturing Systems in Industry 4.0*, EAI/Springer Innovations in Communication and Computing,
https://doi.org/10.1007/978-3-030-90462-3

The manufacturer's authorised representative in the EU is Springer Nature Customer Service Centre GmbH, Europaplatz 3, 69115 Heidelberg, Germany. If you have any concerns regarding our products, please contact ProductSafety@springernature.com

Printed and bound by CPI Group (UK) Ltd, Croydon, CR0 4YY
30/09/2026
02225699-0007